UNPOPULAR Science

An unnatural book about natural phenomena

by

ARTHUR ROSENBLUM

Running Press
Philadelphia, Pennsylvania

Printed in the United States of America

Distributed in Canada by Van Nostrand Reinhold Ltd., Ontario

Library of Congress Catalog Card Number 7484855

ISBN 0914294199

The author gratefully acknowledges permission to reprint excerpts from the following:

Astrological Birth Control © 1974 **Astrology Guide**
The Man Who Reads Nature's Secret Signals © 1969 **National Wildlife** Magazine
Weekly News Report © 1972 John G. Schmitz
Billings Birth Control © 1973 **AP**
Moon Monuments © **Argosy** Magazine
The Magic of Findhorn Gardens by Paul Hawken, as it appeared in **East West Journal**. From the forthcoming book of the same title to be published by Harper & Row, Publishers in 1975.
Chronological Catalog of Reported Lunar Events, 55pp. prepared by **NASA**
Laetrile An Answer to Cancer © 1971 **Prevention** Magazine
Doug Voeks and *Psychic Surgery* © **Bulletin of the California Society for Psychical Study,** Inc.
Sunlight to Power Home © **The Philadelphia Inquirer**, 400 N. Broad Street, Philadelphia, Penna. 19101.
Subscription: 15¢ daily/35¢ Sunday
Nuclear Reactors: Drink Death Pittsburgh © Alternative Features Syndicate
The Positive Effect of Prayer on Plants © 1972 **Psychic Magazine,** Incorporated, 680 Beach Street, San Francisco, California 94109.
Subscription: $5.50 per year.
The Cancer Underground © **Newsweek**, Inc. 1973 reprinted by permission.
Uri Geller: A Step Forward in Understanding by Jerry Shifman © The East Bay Bridge, P.O. Box 1087, Richmond, California 94802.
Subscription: $6.00 per 24 issues.
New Research into Astrological Birth Control reprinted by special permission from **Astrology Guide** Magazine. © 1974 by Sterling's Magazines, Inc., 75 Rockefeller Plaza, New York, N.Y. 10019.
Hypnosis Confirms Fear of Two with UFO Story © Reprinted by permission of **UPI**
America's Largest Nursing School is Now Teaching Psychic Healing © March, 1974, **National Enquirer,** Lantana, Florida.
Material other than introduction and editor's note from
Aquarian Research Foundation

Concept by Lawrence Teacher
Art Direction and Design by Jim Wilson
Back cover illustration by Kent Hendrickson
Type: Times Roman, Composition by Alpha Publications, Philadelphia
Cover printed by Pearl Pressman Liberty, Philadelphia
Unpopular Science printed and bound by R. J. Carroll, Inc., Harrisonburg, Va.

Unpopular Science is published in cooperation with
Aquarian Research Foundation, Philadelphia, Pennsylvania
Art Rosenblum invites readers of **Unpopular Science** *to write or telephone the Aquarian Research Foundation for further information, subscriptions to the newsletter ($10 per year, more or less, depending upon ability to pay), or copies of his book* **The Natural Birth Control Book** *($3.00 Postpaid). The Aquarian Research Foundation is at 5620 Morton Street, Philadelphia, Pennsylvania 19144 (Tel. 215-VI-9-3237 or 215-VI-9-1259).*

Running Press
38 South Nineteenth Street, Philadelphia, Pennsylvania 19103

INDEX

A Word of Explanation

In most books, the table of contents goes in the front and the index is placed at the back. In preparing **Unpopular Science** *for the printer, we thought that it would be much more useful and interesting to break tradition and put a brief alphabetical index up front. We know that placing an index at the front of any book is a most unusual procedure, but because of the rambling nature of the text, this arrangement makes the most sense.*

EDITOR'S NOTE

In the publishing profession, one of the more curious rituals in which writers and editors participate is called the headlining session. For the uninitiated, the headlining session is a brainstorming marathon whose purpose is to come up with appropriate titles, leades, and captions for articles, photographs, or books. As a writer, I learned very early on that headlining sessions were best left to the editors; in fact, I was so hopeless at it that of the 40-odd articles and stories I have had published since 1971, not one bore my own title. As an editor, however, I learned very early on that headlining sessions were best left to the publishers. Otherwise this book might have been titled **A Street Guide to Downtown Atlantis, The Holy Con Man's Catalog,** or **Everything You Ever Wanted to Know About The New Age.**

Although I cheerfully accepted **Unpopular Science** as the book's title (after all, I didn't have a single suitable alternative suggestion of my own to offer), I felt slightly uncomfortable in that Art Rosenblum writes about so much **more** than what I usually understand the word "science" to mean. I would have really preferred to substitute something like "New Age Alternatives," but I suppose that such a phrase is so vague as to be almost meaningless to everyone but myself. Perhaps we can circumvent this semantic difficulty by stretching a point and redefining the word "science" to encompass such things as communal living, group dynamics, and spiritual development.

Unpopular Science is a compendium of some 50 newsletters which Art Rosenblum researched, wrote, published and distributed himself between 1969 and 1974, as part of his Aquarian Research Foundation. The range of subjects touched on is enormous, with everything from astrological birth control to ecological cockroach traps, from ESP in yogurt to mysterious monuments on the moon. For all Art's eclecticism, however, he is often maddingly superficial. The reader who buys this book to find out all there is to know about the aura, acupuncture, or astral travel is very likely to be disappointed. What Art is best at as a New Age journalist is uncovering little-known phenomena, presenting a brief survey or synopsis of the subject, and then giving names, addresses and publications where the reader can find out much more if he wishes.

If I had to state why people should pay $3.95 to purchase **Unpopular Science** when it doesn't even discuss anything in depth, I would probably answer that it is a fascinating browse book that may tell me as much as I really care to know about many subjects I have never heard of. I would also declare that it is an excellent source of basic information about alternative science, medicine, and social interaction. But most important, **Unpopular Science** is a publication that I know will give me access to people, organizations, tools, and above all, ideas. (I must confess to Art that in the two and one-half years in which I have subscribed to the Aquarian Research Foundation newsletter, I have swiped several of his ideas for articles which I later wrote and had published.)

Editing. Because the sheer size of the original material precludes its publication in this book, I had to make many deletions and condensations. I used several criteria in selecting what was to remain and what had to go: all the dated, uninteresting, redundant, or personal references were to be yanked out, but everything pertaining to science (within the scope of my newly construed definition, that is) was to remain. That, of course, immediately presented problems, not the least of which was entire newsletters sometimes had to be edited out. We solved that particular problem by rearranging the book design, dropping the newsletter numbers, juggling the chronological order, and moving some of the later material up front.

Another problem we faced was the wealth of appropiated material Art was accustomed to using in his newsletter. Until we started working on the book, Art was in the habit of using copyrighted material without permission or credit (as many small newsletters and underground publications are accustomed to doing). At one point, we had to check over 40 separate items for copyright permission; that was doubly difficult because in some cases, Art couldn't remember where the material came from, and in other cases, the periodicals had ceased publication. Unfortunately we had to take a hard line: no copyright, no publication. And that wiped out a fair portion of that extra material. Some items, however, have been restored by having Art paraphrase or summarize them in his own words.

Another problem was Art's frequent appeal for contributions, help, and feedback. We can well understand that publishing a newsletter on a shoestring can put one on the razor's edge every month at bill-paying time, and Art is perfectly justified in asking his subscribers for continuing support if they want to receive the next issue. But we also think that to include Art's frequent appeals for money and assistance on top of paying for this book just isn't fair. Hence those deletions. All we will say is that for those who are interested, Aquarian Research Foundation has applied for a tax-free status as an educational institution.

Over the years, a plethora of people have dropped in at the Aquarian Research Foundation to pitch in and help out for a time, and Art has always tried to credit them all in his newsletters. Unfortunately, we didn't have the space, so out they went. Art has asked me to give his heartfelt thanks to all his friends who have visited, helped, participated, or otherwise supported the work these past five years. Thank you, people.

I have, as far as possible, resisted the temptation to edit Art's grammar and syntax, except in cases where it just didn't make sense after several readings. The only major embellishment I found absolutely necessary was punctuating the text with periods, commas, and semi colons. Therefore, if certain passages seem muddled at first, that's just the way Art writes. On the other hand, if there are misspellings and typographical errors, they're the copyeditor's mistakes. What this means, of course, is that I am entirely blameless; it's always somebody else's error.

Another change I have taken the liberty of making throughout the book is capitalizing "New Age." Art was unhappy with this because he feels that the words have been widely misunderstood. By New Age, he simply means, "a new age," and should not be capitalized. I disagreed, I'm editor, and the capitals stand.

Last word. A book is a permanent testimony of a man's talent—or his folly. It is also a chain of creativity that extends from the writer and editor through to the publisher and bookseller. In the final analysis, what a book does for or to its readers determines its worth. It is Running Press' hope that **Unpopular Science** will be as useful to its readers as its ideas may be important to our future.

Daniel Grotta-Kurska

INTRODUCTION

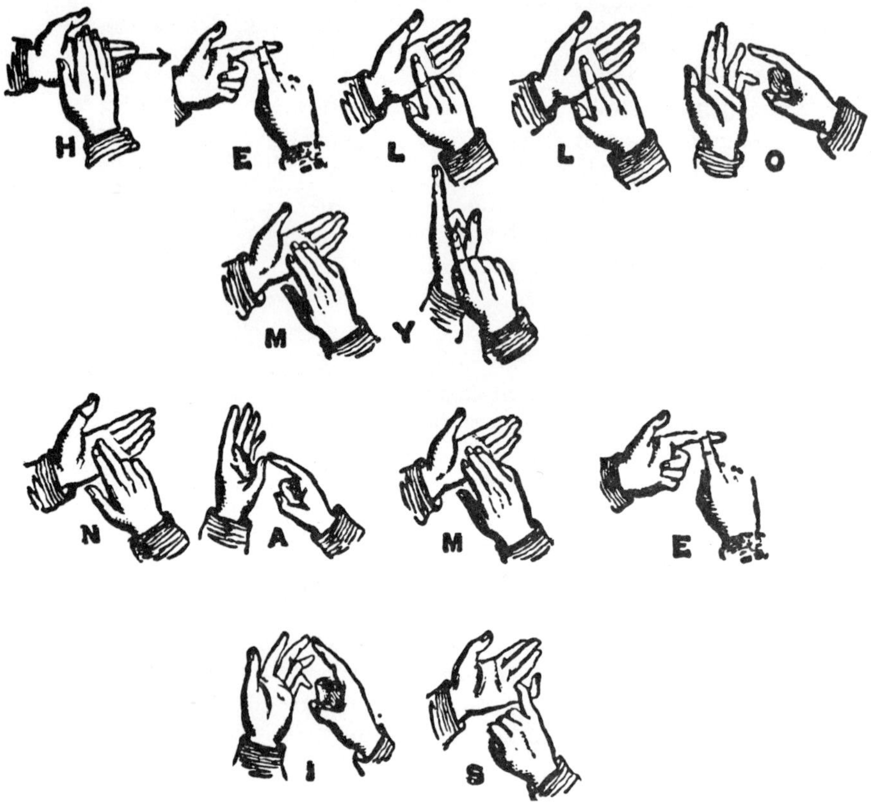

COLIN WILSON

In the spring of 1974, I spent six weeks in Philadelphia, teaching a creative writing course at Rutgers University in nearby New Jersey. When I arrived, the only person I knew in the Philadelphia area was the journalist and novelist Daniel Grotta-Kurska. He had interviewed me in England some months earlier for **OUI** Magazine, and was currently editing a book of my occasional pieces for a New York publisher.

I called to see him at his home in the Mount Airy section of the city, a pleasant neighborhood with tree-lined roads, old houses built in a bewildering but fascinating variety of styles, and a feeling of the colonial Philadelphia of Ben Franklin. As we were reviewing the pieces for the book, Daniel casually mentioned that he had recently written an article for **Today** Magazine, a Sunday supplement to The Philadelphia **Inquirer**, which might provide useful material for a long-planned sequel to another book of mine titled **The Occult**. Grotta-Kurska's article about Art Rosenblum and his Aquarian Research Foundation described him as a visionary with a rather tough-minded, empirical streak. The combination sounded so weird that I was anxious to get a look at this walking anomaly. Daniel was right in thinking that Art might prove interesting, and he later arranged a meeting for us.

Besides, I have to admit, I collect "visionaries." By this I mean that they fascinate me, and I take every opportunity to observe live specimens at close quarters. It never ceases to amaze me that certain human beings can live consistently according to to a set of non-materialistic standards. Ever since I was a child, I have thought of myself as an idealist: that is, a person who is more interested in the world of ideas and imagination than in the more practical pleasures of life. I suppose I became a writer because I wanted to find a way of making a living that would also enable me to pursue ideas. But I never ceased to be aware of the brutal power of materialism and materialists. Hector Hushabye, in George Bernard Shaw's **Heartbreak House**, even suggests that one day the men of ideas may have to rise up and slaughter all the materialists, and there was a period in my mid-teens when that notion struck me as beautiful and inspired. As I wrote my first book **The Outsider** in 1956, I could almost feel Shaw's Boss Mangan breathing down my neck. Now the real visionary idealist is curiously oblivious of the Boss Mangans. The power of his dreams seems to wrap him round like a cloak, insulating him from the world of "practical men." I could mention Michael Byrom, who lives (and paints) in a cottage on the edge of Dartmoor, and who firmly believes that everything to do with the flesh is evil, and that everything to do with the spirit is good. Or William Arkle, the Bristol artist and mystic, whose paintings are full of visionary landscapes that bring to mind the innocence of William Blake's **Book of Thel**. How do these men **do** it? Cynics are inclined to talk about self-deception or plain stupid ity; but my own observation suggests that the visionary is **more** intelligent and honest than the average, not less.

The block on which Art Rosenblum lives had broken glass in the gutters, cracked curbstones, and some singularly dilapidated facades. His own door was painted blue and red and had a yellow peace symbol in the middle, but was otherwise as battered as most of the others. A crowd of black youths eyed me without curiosity as I rang the doorbell.

Art looked much as he does in the photograph on the back of this book: tall, skinny, with a tendency to walk with slightly rounded shoulders. The hatchet-like Jewish profile **does** have a slight resemblance to Abe Lincoln, as Daniel Grotta-Kurska remarked in his article. Art's teeth are large and prominent, and he shows them a great deal, since he usually wears an amiable smile. The photographs I had initially seen in the **Today** article, as a matter of fact, were not too prepossessing. They made him look a weirdo. But when I'd been talking to him for five minutes, I realized that what they missed was the gentle goodness of his personality, the kindly patience that surrounds him like an aura. One critical acquaintance had told me Art talks too much, and seemed to feel that this indicated egoism. Art certainly talks a great deal—if encouraged—but my own impression is that he is almost without any sense of ego. He seems to be an odd combination of a child and a scientist. Everything interests him: the diet of his German shepherd dog (which had just produced about a hundred puppies), the precise meaning of a phrase in the New Testament, the possibility of curing certain diseases with injections of very dilute hydrochloric acid; and, like a child, he wants to tell you about all the things that interest him.

Of course, **anybody** can be interested in health foods, pyramidology, communes, cancer cures, and telepathy; this in itself does not indicate anything out of the ordinary. But in Art Rosenblum, there is a feeling of balance, sanity and a certain dogged **persistence** that makes it clear that he is no crank. He is one of those rare people who has somehow brought together a number of strange and (apparently) contradictory qualities, and I found myself wondering "how he got like he is."

The answer is provided in an eighty page autobiographical manuscript that he lent me, and it is worth summarizing here.

Art's father was a Russian Jew. Tsarist Russia exploded periodically into pogroms; which is why, in 1905, Rosenblum senior's family decided to emigrate to America. He was an idealist, and while a teen-ager in New York City's teeming Lower East Side, he dreamed of becoming a lawyer and helping the workers in their fight against the bosses. Then he met a middle class Jewish girl, fell in love, and decided that it was more important to make a living and raise a family. He went into the mail order business, and achieved so much success that in 1928 (the year after Art was born), the Rosenblum family was able to move from the Bronx to a house at Croton-on-Hudson that he had specially built for them. The prosperity evaporated after the 1929 stockmarket crash; both parents had to work, and Art and his brother and sister were brought up by a German maid named Elizabeth. She seems to have exerted a considerable influence on Art—moreso than his parents. And when, during the Hitler period, he was told that Germans were wicked, he found this impossible to accept, since Elizabeth had struck him as entirely good.

Art's childhood was not easy. He had buck teeth and a speech defect; boys at school jeered at him for being Jewish; he made few friends; he loved the countryside, and hated it when the family had to move back to the city. His mother discouraged him from playing with non-Jewish children —which seems to have been ill-advised—but there were few Jewish children to play with. He was not particularly close to anyone, not even to his brother and sister.

It was probably the loneliness and isolation of his childhood that led to his later obsession with communes and groups. His idealism is strongly people-oriented. The Jews have a powerful Utopian tradition, a dream of the City of God; in Art, it was given force and direction by the longing to lead a normal, fulfilled social life. He describes a recurrent dream of his childhood in which children, standing hand-in-hand, shouted to him: "Arthur, come, ARTHUR, COME. . ." He even went cycling beyond the hills that were visible from his bedroom window, hoping to find them. He also describes how, at the age of eight, he stopped to stare at a group of naked children who were dancing in the spray from a garden hose. One of them invited him to join them, but he was embarrassed and refused. On his way home, later, he decided to join them if they were still there; but they had gone. The incident is somehow symbolic. The dream of joining a community of happy, uninhibited, creative human beings has persisted throughout his life.

At the age of 14, the longing expressed itself in the form of a vision. Walking up a hill, he suddenly had a sensation of floating in the air and looking down on a group of young people—which included himself—working on the roof of a house. He knew they had built the house together. There was a sensation of freedom and joy. The longing for community was expressing itself as a definite sense of purpose.

In another similar vision, about a year later, he saw himself as being ten years older—in his mid-twenties—and again a member of a happy, united group, who not only worked together, but enjoyed healthy, beautiful relationships. But as he grew older, the visions began to fade, and finally, disappear.

Art's parents were Zionists, and through them he heard about the Kibbutzim in Palestine. For a while, this seemed to be the answer he was looking for. Daniel, a friend of his brother's, was an active Zionist, trying to recruit young Jews for Kibbutzim. Art asked if there were community groups who were pacifists, determined never to fight or kill under any circumstances. Daniel investigated, and discovered there were no such groups at the time. Daniel also arranged for Art to go out on a date with a Jewish girl—another group member. She intimated that she would be happy to go on dating Art—perhaps even to finally marry him—if he became a member. It was a temptation, since he was—and is—deeply interested in sex; but his pacifism won out. Reluctantly, he dissassociated himself from the group.

This story illuminates an important aspect of Art Rosenblum's character. The longing for community is powerful; but the idealism is intransigent. This is what distinguished his Aquarian Research Foundation from so many similar organizations concerned with aspects of "the unexplained." By using his Newsletters simply as a platform for weird beliefs, he could no doubt amass an army of disciples. But he is concerned to reach intelligent, sceptical people, who are **not** willing to believe "ten unbelievable things before breakfast." Hence the empiricism, the scepticism, of the approach that can be observed in these Newsletters.

The scientific interest seems to have developed in his middle teens. He made friends with an expert in electronics, who fired his imagination with talk of scientific research. Art was shocked to learn that many important inventions were bought up by large manufacturers, and simply left unused, to keep them off the market. These people regarded an invention—a practical distillation of human knowledge—as a commodity that could be used in war or money-making. For Art, the knowledge was an end of itself. Like his father, he was developing into a radical as well as a visionary.

Art went to work in an electronics laboratory at age 16, but got fired in the first week for arguing with the boss. They took him back—and immediately fired him again. He was definitely not a conformist.

This also became clear when he went for his Selective Service physical in 1945. The doctor asked him what he would do if a German was about to shoot his mother or sister. Art, suddenly deeply moved by conviction, said "I'd just as soon shoot you as a German." The doctor responded angrily, and it struck Art that if they were going to yell at him like this, he wouldn't be able to stick the army. Fortunately, his papers were stamped rejected.

There was a period as an engineering student at Wayne University, which he hated, then at Hiram College, Ohio, which he loved. There he became acquainted with the Quakers. One of them told him he intended to go to join a Bruderhof community in Paraguay, where everyone worked and shared all they had, and where all were opposed to war. A pamphlet told him that the Bruderhof communities had been started in Germany in the 1920s, and that this one had been forced to leave Germany by Hitler. Art immediately decided to join. He wrote to them, and received an encouraging reply. He raised the boat fare—$275—and sailed from New Orleans. The community was a day's journey from the capital city, Asuncion; it consisted of three villages and about 600 people. Art was dismayed to learn that it was a **religious** community—something the pamphlet had failed to mention. But having come so far, it seemed a pity to tell them he was an agnostic. (He did, later.)

Although Art spent fifteen years with the community, he was always something of a misfit. To begin with, there was never any free discussion of sex. Their attitude towards it was distinctly puritanical. Now sex had always been of considerable importance to Art. (The researches of the American psychologist Abraham Maslow revealed that nearly all highly dominant human beings are highly sexed; Art, is his quiet way, undoubtedly belongs to the dominant group.) Somehow, sex was included in his vison of Utopia. Instinctively, he agreed with the view Blake put forward in **Visions of the Daughters of Albion**, of a guilt-free, jealousy-free sex. (He would probably have found himself at home in the original 19th century Oneida Community, established in New York State by John Humphrey Noyes, with its concept of group-marriage.) He felt cut off from the other young people of the community by his guilt about the natural desires he felt for the girls, and the part they played in his sexual fantasies. When he asked the brothers why they would not allow free discussion of sex, they replied that when young men and women discuss sex, they are not far from doing it. This struck Art as an unreasonable point of view, but he pretended to agree; if he hadn't, it would probably have been impossible to remain. He hoped to marry in the community, but somehow this never came about.

Eventually, he became a "Believer," and it happened in a peculiar manner. He had told the brothers that if God really existed, and wanted him to believe, no doubt He would make Himself felt. Later, Art's family sent him the money to return to the States to visit them, and he also hitchhiked around for some months, performing various errands for the community. Odd "coincidences" began to occur that made him feel that human life is not shaped purely by chance. For example, he was anxious to see the Dean of Hiram College, to tell him about the community. On his way there, he found it difficult to get lifts—until the Dean himself stopped and offered him a ride. The Dean had intended to stay home that Sunday afternoon, and he suddenly felt impelled to go out for a drive to nowhere in particular.

It might be objected that this kind of experience neither proves nor disproves God, although it may prove telepathy or other froms of extra-sensory perception. But when an atheist says he doesn't believe in God, he also means that he believes life is basically governed by chance, and that religions are forms of self-deception and wishful thinking. As you look around at the everyday world, there is little sign of a God. But if odd coincidences start to happen, you get the feeling that something, or someone, **is** trying to tell you something. In my preface to **The Occult**, I described how such "coincidences" happened all the time during the writing of the book—how, for example, I one day needed to find a particular reference, with no idea of where to start looking, and as I idly pulled a book off the shelf, the volume I needed tilted off the shelf, and fell open at the right page. . . . Such"luck"is usually associated with states of optimism and vital intensity, and may be due to some form of mental radar. But it certainly suggests that the "accident and chaos" theory of the universe may be only the result of the personal shortcomings of the "scientist."

On this trip to the States, and on subsequent visits, Art continued to visit other similar communities. This notion of community is one of his basic permanent interests. What is very clear from his descriptions is that such communities never quite corresponded to his mental ideal. He found himself wondering at one point why, if this communal life was really like early Christianity, no one had the miraculous gifts of the apostles. The answer, clearly, is that community is not enough. All the great advances in the history of man have come from inspired **individuals**; and because the apostles were inspired by Jesus, their powers were not passed on, but died with them.

At all events, Art found himself increasingly dissatisfied by the life. A girl he was hoping to marry refused him. The Brothers said he seemed to be too critical, unable to see his own faults, and for a while he was informally "expelled" until he had got rid of his pride. . . He spent a year as a customs despatcher in Asuncion, still working for the community, and enjoyed the experience.

Finally came the parting of the ways. This was in 1964, shortly after the community moved to the United States. Art felt the need for some more active involvement in life. He left the Bruderhof and join the Civil Rights movement, living for six months in a religious commune in Atlanta, Georgia. This was the beginning of a five-year period of wandering and "finding himself." He worked for Students for a Democratic Society (SDS) for a year, as their printer. He travelled around helping other civil rights and peace groups to set up printing presses. He moved from one commune to another. One of the most interesting, from his point of view, was the Fort Hill Commune in Boston, where there was no objection to males and females sleeping together without being married, and where, if the girl became pregnant, she could feel certain that her baby would be cared for as part of the community. This was obviously closer to the ideal he had been searching for, and Art got over many of his life-long sexual neuroses. Yet all this experience of communities, peace groups and religious societies only confirmed his basic feeling that **he himself** had to find the way to do something positive and creative.

The problem, he felt, was that left-wing protest groups may be sincere, but that they nevertheless do more harm than good in trying to convey their views to ordinary people. For better or worse—probably better—most people are rather conservative, and they distrust revolutionaries. Militant beards and sandals only alienate them. Art felt that the chief problem is to somehow **reduce the resistance** to new ideas; to change. This seems to have been the germ of the idea that, in 1969, developed into the Aquarian Research Foundation, based in Philadelphia, (perhaps chosen by Art because of its connotations as the "city of brotherly love.")

Art had come to realize that science has to be somehow persuaded to widen its boundaries. The idea came to him at an excellent time—when it had started to happen anyway. In the 19th century, science was almost a religion of "tough mindedness"; it seemed to take a grim pleasure in pointing out that few of man's beliefs about himself and the world are really tenable. This led to the view —expressed, for example, in Walter Stace's important book **Man Against Darkness**—that man is a biological accident in a meaningless universe, and that he had better make the best of this. Freud talked with a shudder about the "black tide of occultism"—meaning self-delusion.

But then, in the 1950s and 60s, some of the basic attitudes of science began to change, due largely to the efforts of men like Karl Popper and Michael Polanyi. Their point was that scientific discovery is a process as "inspirational" as the writing of poetry, and that the scientist who pretends he is somehow the **opposite** of the poet is a liar—either that, or a poor scientist. "Reductionism" had been the fundamental creed of science—the attempt to explain things in terms that can be translated into a well-designed experiment. Freud's essays on Leonardo and Dostoevsky may be taken as fair examples of reductionism. The non-reductionist points out that if you "explain" Dostoevsky in terms of an Oedipus complex, you are ignoring the most important part of him; the Freudian theory is like a net whose meshes are so wide that it is only good for catching whales. The reductionist reply would be either that the smaller fish do not exist, or that they are so unimportant that they don't matter. The one thing he is totally unwilling to admit is that there is something wrong with his net. Abraham Maslow is a good example of a psychologist who was repelled by reductionism, and who tried to broaden psychology until it **could** take into account man's creative and mystical impulses. His concepts, such as the hierarchy of needs (which says that we all evolve up a kind of ladder of needs, starting with food and shelter, then going on to sex, love, self-esteem and, finally creativeness) produced a new synthesis of Marx, Freud, Adler and Jung.

Back in the 1930s, J. B. Rhine set up his Parapsychological Institute at Duke University, and made some attempt to investigate ESP phenomena scientifically. His experiments with cards, dice etc. (testing whether correct guesses could be explained strictly in terms of chance) were classically designed. But, understandably, they failed to **expand** the field of research. They seemed to prove very little except what they set out to prove—that telepathy existed—and even there, there were doubters. Polanyi and Popper could have told him the reason. Science begins with inspired guesses, then sets out to test them. The imagination creates some apparently absurd hypothesis in the same way that J.R.R. Tolkien created Middle Earth, and then sets out to explore the new territory. As sub-atomic nuclear research into such strange particles as quarks has shown in recent years, there is no harm at all in a scientist seeing how many unbelievable things he can believe before breakfast—provided he then sets out to test them. In this field, an excitable imagination is more than an advantage—it is a necessity. But then, the imagination feeds on **possibilities**; they provide the stimulus, the excitement. Occasionally a purely "scientific" experiment may open exciting vistas—as when Cleve Backster decided to attach a lie detector to a philodendron plant in his office, then see whether it would react when he burned it with a cigarette lighter. He seems to have demonstrated, fairly conclusively, that plants can detect our thoughts. On the other hand, devices like Kirlian photography and the tobascope—both of which seem to be able to record the "life force" that radiates from us—could not have been invented unless scientists had been thinking in terms of life-fields, auras and other concepts that would have driven Freud to profanity.

It is easy to understand Freud's detestation of the "black tide of occultism." When he was a young man, doing his early work on hysteria, Europe and America were in the throes of an "occult revival." (I say a revival because something very similar had happened towards the end of the previous century.) Periodically, its leading figures—like Florence Cook and Madame Blavatsky—were exposed as frauds; but this made no difference to the enthusiasm of believers. Naturally, Freud found it easy to believe that this was another manifestation of hysteria and neurosis. Besides, it must be admitted that there is something about spiritualism that makes healthy people draw back with a slight wrinkling of the nose. Robert Browning portrayed Daniel Dunglas Home as the fraudulent medium Mr. Sludge, although in fact, Home was almost certainly genuine. Nathaniel Hawthorne witnessed Home's ability to make heavy tables float and wash his face in blazing coals, and admitted that it was genuine—**and** oddly boring and irrelevant. There is a great deal of evidence to suggest that there **is** life after death and a world of spirits. The trouble is that, as often as not, spiritualism seems to attract the wrong kind of people. All this explains why most scientists and philosophers decline to take any interest in the findings of the Society for Psychical Research, even when they have been arrived at by the most strictly scientific methods.

Now the present occult revival —which began in the mid-sixties —also involves a great deal of crankery and gullibility. But there is far more to it than that. Dr. Harold Saxton Burr of Yale University attached delicate voltmeters to trees, and established that the tree's life-field fluctuates with sunspots and with the seasons, that a woman's ovulation can be detected by the same means, and that cancer patients show a characteristically "unhealthy" field. The phenomenon of hypnotism —described for example, by Dr. Stephen Black in **Mind and Body**—indicates that the mind has a far greater control over the body than medicine has ever realized. A statistical survey of successful men reveals that there is some odd correspondence between their careers and their astrological signs. (Michael Gauquelin has discussed this in **The Cosmic Clocks**.) What is happening is that certain scientists are beginning to take an interest in these problems—an interest that is free of the slightly morbid, cranky touch that used to characterize "occultism." Freud disliked occultism because he felt that "occultists" wanted to drag science back into their shadow-world. But the aim of investigators like Harold Burr, Cleve Backster, or the Soviet scientist Victor Adamenko, is to carry the floodlights of science **into** the shadow world.

And this explains why I find Art Rosenblum such an interesting and important figure. His mind is sceptical and scientific, but he is also "wide open," willing to investigate anything that looks interesting. He says in one of the Newsletters: "The work of the Aquarian Research Foundation has never been predetermined in the kind of logical way that most institutions are operated. Our plan has been to use intuitive ability (or openness to the leading of the Spirit) as a means of deciding what to do." I found it at first difficult to adjust to this mixture of openness and scepticism. I asked him what he thought about the new belief that pyramids have some odd preservative effect—as explained, for example, in Lyell Watson's excellent book titled **Supernature**. (Watson also claims that his razor blade stays sharp if kept in an east-west alignment in a pyramid; Marshall McLuhan's son Eric has made the same kind of claims, although he says it doesn't work on stainless steel.) Art said he had obtained plexiglass pyramids of the correct dimensions, and carefully performed various experiments to determine whether (a) meat placed in the middle of the pyramid would "mummify" rather than decay, (b) whether seeds would grow faster inside a pyramid, (c) whether certain colloids would "thicken" more quickly in a pyramid. He concluded that the pyramid simply did not work. He proposes to repeat the experiments more carefully, to see if the negative result is due to experimental error; but at present, he is a total sceptic about the powers of pyramids.

On the other hand, a visit to Cleve Backster's laboratory convinced him that plants **can** sense our thoughts. He told me that Backster had "introduced" him to a plant, and then, to test whether the plant had "established sympathy" for him, he sprayed Art's hand with a local anaesthetic which froze the skin; the plant registered Art's pain in the graph. Art was then told to take two stop watches, and to go and urinate in the toilet. He was to stop one watch as he began to urinate, and stop the other as he finished. He was then to flush the toilet. The flood of disinfectant would destroy the living cells in his urine, and the plant should register their destruction "sympathetically." Art prepared to urinate, stopped the watch—then found he had to wait another ten seconds before anything came. He stopped the other watch as he finished urinating, but as he went back into the laboratory, remembered that he had forgotten to flush the toilet. Backster looked at the stop watches, then studied the graphs obtained from the lie detector. He looked puzzled. Art told me the graph showed a regular wave motion, about one per second, which then went into longer and slower waves as he started to urinate. Perhaps of contenment? When he stopped, the small one-per-second waves began again. Backster finally said: "For some reason, you didn't start urinating until ten seconds after you stopped this watch—as shown by these ten waves after the time you stopped the watch—and you also forgot to flush the toilet. Where were you brough up, a barn?" he kidded Art. Art blushed, and the pen on the graph went wild, making violent up-and-down sweeps. "You see," said Backster, "That's your embarrassment. . ." If Art had flushed the toilet, the plant would have registered the death of the micro-organisms in a similar way.

From all this, I got the impression of a hard-headed sceptic. (He hates to be called an "occultist," or to have the word "occult" used in connection with his work.) Yet a moment later, he was telling me that shortly after establishing the Aquarian Research Foundation, he decided that he ought to ask the spirits for their advice. The famous English medium Edna Cox told him that he didn't need to go through her; he could consult them directly himself. He was to put a handwritten message on his bedroom wall asking the spirits to wake him at precisely 6 A.M. He did this, and woke at six exactly. After this, a series of vivid dreams told him what he needed to know about the right approach. In another incident involving mediumistic advice (this time the Vietnam war,) he actually got answers from influential people in Congress, and believes it is possible he may have influenced the course of events that led to the end of the war.

All of which should make it clear that Art Rosenblum is an interesting combination of scientist and. . . perhaps the right word is **mystic**. He believes in the "unseen world," and he wants to investigate it scientifically. He is a kind of liaison man between the world of people with unusual claims, and the people who want to know if they are true. He is essentially a catalytic force.

And it strikes me that now, in his mid-forties, he is at last beginning to realize the dream of his early teens—that dream of happy people collaborating on building a house. His life in a commune was no answer; the community spirit was there, but they were not building a house—at least, not Art's kind of house. It is significant that he came closest to real self-fulfillment in learning German and translating the life of the healer and spiritual leader Johann Christoph Blumhardt. He has had to learn what all creative thinkers have to learn: that—as Shaw put in in **Back to Methuselah**—"You can create nothing but yourself." Although he has helpers, he **is** the Aquarian Research Foundation. Now he no longer lives in a large community; much of his time is spent alone. But, with his Newsletters and his work, he is establishing a real community, a community of sympathetic and intelligent human beings, all concerned with the evolution of the human race.

These Newsletters are full of fascinating material, from cockroach traps to cures for cancer. There is material here to keep Art and his helpers busy for the next ten years. As I read through them, I have an exciting feeling that here is the spirit and force to make a new revolution. The Foundation is young, and Art Rosenblum himself is a comparatively young man. I have a feeling that I have stumbled upon something that may exercise a wide influence, and that Art Rosenblum's real achievements still lie in the future.

Colin Wilson
Gorran Haven, Cornwall
May 1974.

AQUARIAN RESEARCH FOUNDATION

Our purpose at Aquarian Research Foundation is to do research to find ways whereby the New Age can come in a beautiful way of love instead of through a time of chaos or civil war. One of our methods of supporting ourselves while doing this may be an educational project. But since our concern is to tackle the established society and find out how to break down its resistance to change instead of running a school for young people, we might try to give courses for parents of young radicals to help them to understand why the kids are thinking and doing the things that they are. If this is possible, we ourselves may be learning how to break down the parents' resistance to the changes that their young people are experiencing, and this will give us clues about breaking down the resistance of the larger society also. But, of course, many other methods will also be sought and experimented with, and I am confident that one thing will lead to another.

It may be that some of you have been thinking about this project and would like to come and visit us. That is possible. There is also no intention to limit the size of the group. There are even more houses like ours on the same block in case this one gets filled up.

PUPPIES: We also have a beautiful girl dog named Maudy, who is half collie, half shepherd (but she needs a license). We want to fix her a box by the front window so the children can look in and watch the pups she will be having shortly. If you need a puppy check in early. Reservations are in order.

BLOCK RADIO STATIONS: While I am typing this, I would also like to be able to start trying out our "block radio station," which will work on the regular broadcast band by carrier current so that it is legal without a license. We hope that this will be able to help us relate to the people on this block, and help the young people here relate to their own parents, who will probably listen to them on the radio, even if they don't in the living room.

MORE PUPPIES: Maudı has given birth to eleven healthy puppies around 3:00 A.M., November 4, 1969. Mother and babes all doing well. They will make fine anti-establishment Christmas presents as they will be ready for delivery to their new masters about the 17th of December. They are 75% German shepherd and 25% collie, which makes them sharp watch dogs and lovable companions with a character—judging from Maudi—as close to human as you can get. In all probability, Maudi will then return to our friend, Germa (her owner), and we will keep one of the pups.

SCHOOL PROJECT: We recently had a little meeting and talked about the procedure for organizing and starting the classes for parents. Plan is to charge each family $5 for the evening, with some allowance if people are hard up. We also spoke of having a public meeting here every Thursday evening with a lecture on "What is the Age of Aquarius?" followed by a discussion, and to ask a donation of $1 to $5 per person, according to ability. People who come will also be invited to help with the house if and when they can. We are also planning to invite Quakers to a "pot luck supper" with a discussion on the same subject.

BLIND READER: Also, last night I had a call from a girl in New Haven who has been thinking of joining me in this work for some months now. She is thinking of quitting her job and coming in four to six weeks, but wants my reaction first to a tape-letter she is sending me. She is blind, and I think she could be a great asset to the group. I told her about an idea I had for a device which would enable blind people to read ordinary print, and which I hope to talk over with the director of a company called "Bionic Instruments," which does non-military elec-

BLOCK RADIO STATIONS

A block radio station is a radio station that operates on the standard broadcast band (frequency 640 K.C.) and covers an area of from one to four city blocks. It can be operated legally without a license (under Section 15 of FCC regulations regarding carrier current transmission). However, if the operation of the station leads to complaints from people in the neighborhood, it may be closed down and even possibly that the equipment will be confiscated. If the station is used in such a way as to get the general approval of everyone in the neighborhood, there should be no trouble. It cannot be used for anti-government political programs, and FCC regulations against obscenity must be observed.

The main purpose in setting up block radio stations would be to give young people and others in the community a chance to put their own music and programs on the air. It is a way to encourage parents and young people to start listening to one another. If the neighbors are able to hear what the young people are saying, and some younger brothers and sisters are listening also, it is quite likely that parents will also want to listen. As that develops, the parents will come to understand more of the ideas, concerns, and problems of the younger people. At all times, care would be taken to keep the station in a positive light in the neighborhood and set a good precedent.

The equipment required will be a record player, one or more good quality microphones (dynamic), a tape recorder with automatic gain control, and an earphone or monitor output jack (such as the $40 Concord F-50), and some switching arrangements which local technicians can design.

All input of speech or music or a connection to the phone line (now legal for "talk shows") would be fed through the proper resistors and capacitors to the microphone input of the tape recorder, which would be set in the "record" position (though the tape need not be running). The monitor or earphone output of the recorder would feed the signal into the transmitter audio input, which is adjusted so that the correct level of modulation is automatically maintained by automatic gain circuit of the tape recorder, regardless of how loud people speak.

The transmitter should have three to four watts of R.F. output power, which is not fed to an antenna, but back into the power line in that area. No permission is required from the power company. Theoretically, the area covered would not exceed that which is fed by the same power transformer which feeds the transmitter, and would be limited to one block or

Continued →

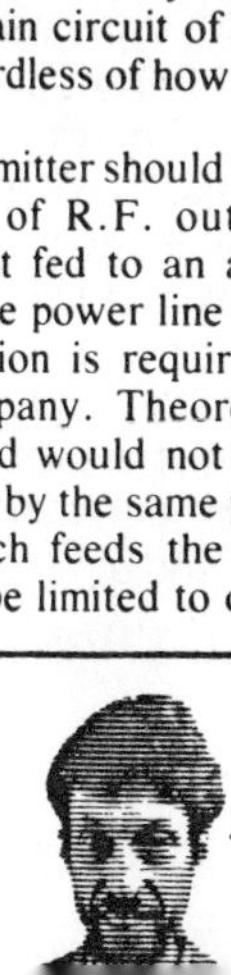

less. In practice, however, it is often possible to get a clear signal two blocks away from the transmitter in some directions; it may even cover four square blocks. I have not tried the system in rural areas.

The transmitting equipment used is the same as the smallest unit commonly used for college campus radio stations, and is supplied by the Low Power Broadcast Company, 520 Lincoln Highway, Frazer, Pennsylvania 19355, (215) 644-4096. It is known as the RC-6a, four-watt transmitter, and employs a toroid matching unit to couple the R.F. output to the power line. The transmitter output may be connected to the same power cord that feeds the transmitter. Reverse connections to test for best polarity.

The transmitter costs about $200, but is well worth it as it has good modulation characteristics and couples well to almost any line conditions. It is also very reliable and trouble-free. Building your own equipment for less money is possible, but according to FCC regulations, it has to be of an approved type, so that home-made equipment may not be satisfactory if any questions were asked by the authorities. Solid state transmitters were also used experimentally, but are not easily protected from line surges, etc.

At one time people had the idea of linking together a larger number of block radio stations to cover a wide area, but it never got going. Doing that would also defeat the ideal of a local grass-roots control. However, tapes of good programs could be copied and exchanged between various stations in any area.

In some places, it might be possible to use already existing campus radio stations and feed the programs into the powerline outside the campus, but I have no experience in this. The main thing would be to be certain that too much power was not radiating. With a properly operating transmitter, a moderately good transistor radio (eight transistor) should lose the signal when it is across the street and half a block down from the house in which the transmitter is feeding the powerline. But when the radio is held against a powerline entering a house it should pick up easily.

tronic research. My idea is to design a computer, with separate elements that could scan different type faces and then "read" the reflected light off each letter, and then it would produce a different tone for each letter. The blind person would have to be trained to recognize the tones, of course. The Director also offered to give me any help I needed to get the radio transmitter working properly. It only remains for me to find the time to go out there.

COMMUNAL LIVING GROUPS: I get a lot of letters asking for the addresses of other communal living groups and wish to report the "The School of Living," Heathcote Center, Rt. 1, Box 129 Freeland, Maryland 21053, is presently compiling a revised list. Send 25¢ for that.

Last night I had a call as a result of our ad in the **Drummer**, from Frank Genco, who is a professional hypnotist. He tells me that he would like to help with our work and that many of the students and young people to whom he has taught hypnosis are also interested in doing something further, and he would be interested in helping us to get in touch with them. He invited me to a class of his and offered to give classes here at the Foundation. Many people are unaware that hypnosis, properly handled, can have all the beneficial and liberating effects that are sometimes obtained from drugs, but without the illegality or other dangers that drugs involve. Hypnosis cannot force people to do things (like commit murder) if those things are actually against their will. It is used by many psychiatrists to overcome deep-seated personal problems. A great deal of new information can be obtained by that means, but the information so obtained must be carefully checked.

The parents class has not been as easy to get started as I had hoped. A short article did appear in a major newspaper, but had little result, mainly, I suppose, because it appeared too late and we did not have a telephone listed under the name, Aquarian Research Foundation.

There now appears to be a good likelihood of getting the parents class started, but it will take a lot of work. Several Jewish and some church groups are interested in helping, and we are now listed with the Philadelphia Community Referral Service of the Health Department.

WFNRS, our Free Neighborhood Radio Station, is off the ground and on the air! We have been broadcasting for five days now from 4 to 8 P.M. each day. Being unable to get our homemade radio transmitter working properly, we visited Dick Crompton of the Low Power Broadcast Company (520 Lincoln Highway, Frazer, Pennsylvania 19355) and told him what we were trying .o do and asked for his help. After reading some of our articles, he came up with a very slightly used 3 watt carrier current transmitter (such as many college stations use) and let us have it for half price. ($92).

This transmitter works perfectly. We also bought two cassette tape recorders (a Concord F-50, which was recommended, and a Nivico, which was cheaper). Both of these have automatic recording level control. One of them is being used as part of our operating equipment so that the transmitter is always working at full modulation without distortion, regardless of how people speak. This is necessary with children. We are also setting up an arrangement to put phone calls on the air.

Our greatest need with the radio station is to have a person here who can relate to the neighborhood and find ways of making the radio station relevant to them. One thirteen year-old has offered to be our disc-jockey for 3 hours each week.

Paula Buntele, who has been here for six weeks now, has been operating the station and manages well. She has the advantage of being able to borrow all sorts of books and literature prerecorded on records and tapes from the library for the blind, and we have been broadcasting C.S. Lewis' "Narnia" stories, which are highly imaginative tales of another world in which several levels of understanding can be found. Later we can present **The Hobbit** and the whole Tolkien Trilogy.

Station Coverage: We have not yet got a full picture of the total area covered by the station, but it is surely more than a square block; possibly 3 or 4. We need someone who can visit and canvas the neighborhood.

Cost: A transmitter costs about $200. Tape recorders, turntable, etc., cost another $150 to $200. No license is required.

Call letters: Any five letter call may be used.

WHAT THE ARF IS DOING.

Many people are able to see that this country is heading for a civil war instead of entering into the spirit of the New Age. Ours is becoming an age of violence, not because of the coming New Age, but because of society's resistance to change. The purpose of starting this community is to find answers on how to break down society's resistance to change.

Methods of Research: It seems apparent that resistance to change is caused by fear, and that fear is mainly due to ignorance. New and extremely rapid ways of overcoming that ignorance must be found. The present experiment of setting up educational block radio stations through which all the people in a small section of a city can begin to talk and listen to one another and share their ideas, concerns, and problems might be a beginning. The classes for parents of radicals, for which there is now some solid hope of sponsorship by a well-known church, might also be a method. But these projects which we are already starting are only a small beginning. To change the whole future of a nation, and possibly the world, **is a tremendous task and calls for a completely different way of doing things than any institution has ever attempted. Few of us yet have the background for it. Therefore, it is not easy for us to understand.**

Science, which has evolved in the last few centuries, has shown mankind a whole new method of overcoming ignorance. But science has been terribly misused.

An equivalent breakthrough, much faster, and led by a spirit of love, must now come about if the responsibility of science is to be corrected in time to avert a catastrophe.

The experiences of Johann Christoph Blumhardt (a 19th century German pastor who is as yet almost unknown in America) and the experiences of famous psychics such as Edgar Cayce have proven that such new breakthroughs are still possible. The early Quakers also showed it. By opening themselves to listen and be used by the Spirit that enlightens every person, new insights could come to a group of seekers. Sometimes these insights are sudden. They can also lead to united action.

Communal societies that are somewhat successful (most of which are still unknown to the public) have also shown that it is possible (with effort and struggle) for people to learn to listen to one another, share openly their deepest fears and longings, and "get a-top of" many problems that have no immediate solution. They can become so free of fear that they can act upon the new insights that may come to them.

The surest way to discover new methods of changing the world, or yourself (or, in fact, both at once) is to **act upon the deepest insights that you have already been given. And that, finally, will have to become our method of research. That is how we have to find out how to break down society's resistance to change.** If it succeeds, the chaos we are heading for could be a chaos of freedom, joy, humor, and much love. If it fails, and changes that the New Age requires do not come about, we might expect the world's most horrible mess, and all the preparations for that have already been made.

In the last few weeks, we have had a large number of requests for information and quite a few visits and meetings. A number of people have also expressed a wish to join fully in the work of the Foundation, and a few have visited with that in mind.

We found that very few people have enough background to understand what we are trying to do and how we hope to do it. The area which is **least understood**, and which is almost entirely omitted from the educational systems, seems to be the very important one **of the meaning and purpose of human life and what the New Age is all about.**

Of the tremendous amount written in these fields, most is philosophy and conjecture—very little being factual. But in recent years, some very solid experiences and factual studies have also been reported, and we came to feel that it is important for people to know about these before starting to work on this project by becoming part of our community. Therefore we made a list of some such books and have started to ask people who want to join us to read these books before making a final decision to come and work with us.

A number of books are:
Breakthrough to Creativity by S. Karagulla, MD, published by De Vorssl Co., 4900 Eagle Rock Blvd., Los Angeles, California 90041 $5.95

Edgar Cayce, The Sleeping Prophet by Jess Stearn. Bantam Books, Paperback. $.95.
What I Believe by Edgar Cayce. Published by A.R.E. Press, Virginia Beach, Virginia Paper $1.25.
The Other Side by James Pike, published by Doubleday & Co., about $6. Libraries have it. It is now available in paperback.
The Morning of the Magicians by Pauwels & Bergier. Avon Books Paperback $.95.
The Kingdom and the Power (has some good things and some doubtful parts.) by Louise de Wetter Published by Christian Ravndal, Thomas Road, Rindge, New Hampshire 03461. Paperbound $1.65.
The Life of Johann Christoph Blumhardt by Friedrich Zundel. A summarized translation by Art Rosenblum (translator's notes added). Not published. Copies exist at ARF and may be printed sometime in the future.

NATURAL CURE OF INFECTIONS

We have received information on the natural cure of infections from an American doctor in New Zealand, and have had two opportunities to try at least part of the method with outstanding success. In each case, the infection happened to be one of the inner ear of a young woman.

1) was that of a young person who was here for a little over a week when she announced that she had an ear infection which had been developing for that time, and was now so bad that it hurt when she moved her jaw. We rushed her to the hospital, but they refused to look at her because her parent was not present to give permission, and we were not her legal guardians.

Following that, we called a medical doctor, but found him to be out of town. It was Saturday night, so we administered the following according to the suggestions of Dr. Conroy (New Zealand). He told us that he got the hints from Dr. Klenner, chief of staff of Memorial Hospital in Reidsville, North Carolina. We have not yet had time to check with Dr. Klenner.

We administered 15 to 20 grams (not milligrams) of natural vitamin C and two teaspoons of powdered garlic, and told the girl that if there was no definite improvement by morning we would have to telephone her parents. Pain was considerably less by morning, and a few days of similar treatment cured the matter completely.

2) was a girl who had had an inner ear infection for over a month. She was receiving antibiotics from a doctor, but on the day we saw her, she told us the infection was getting worse. We told her our previous experience, and she took natural vitamin C and garlic capsules (from health food store), and the infection went away in a very few days.

For serious infections, Dr. Conroy suggests 15 grams vitamin C (which may be part natural, part ascorbic acid, or all natural).

Additionally, six to ten garlic capsules and two to three teaspoons of cider vinegar. Also, fasting for several days, taking only juices and broth. All the above is taken daily.

SPIRITUAL COMMUNITY GUIDE We have now received our copy of the **Spiritual Community Guide**. The publishers, "The Spiritual Community," Box 1080, San Rafael, California 94902, hope to publish additional material of a similar nature, including an international guide to spiritual communities and a newspaper which will update the guide continuously. They ask readers to send suggestions, names of spiritual groups, addresses, reports, etc. They use the word "spiritual" in the broadest sense to include yoga, natural food sources, communes, etc.—any group concerned with things spiritual.

The cost of the guide is $2.95.

HYPERACTIVE CHILDREN: There may be various reasons for the rise of hyperactivity in children (which has become such a problem in schools that kids are being drugged). One possible cause seems to be radiations, possibly in the very soft X-ray spectrum.

The Environmental Health and Light Research Institute in Sarasota has been doing experiments along this line and found a "very significant calming down of hyperactive children and notable improvement in their ability to concentrate on their studies and even several extreme learning disability cases very quickly learned to read. . ." People wishing to follow up on this could contact Dr. Lewis W. Mayron, 5437 Suffield Terrace, Skokie, Illinois 60076, as he is in charge of the research.

The changes they are experimenting with are in the direction of producing a natural light environment equivalent to sunlight (with natural proportions of UV light) and a total intensity of 75 to 100 foot candles. Lights are shielded to stop X-ray emission.

At ARF, we are converting to that type of lighting.

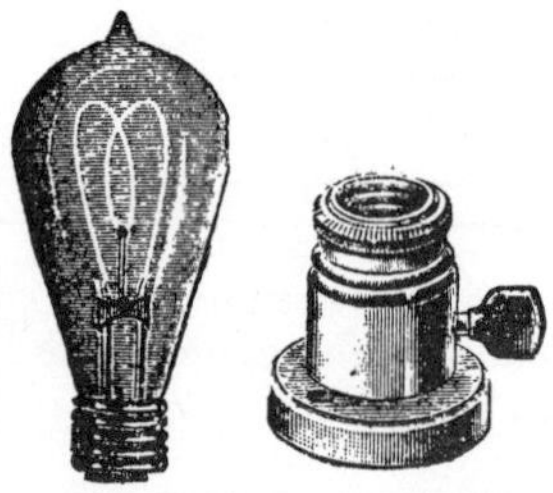

NATURAL LIGHTING: Though it seems a bit commercial, we decided to tell you about Vita-lite fluorescents, which are the outcome of the research described above. Durotest, who manufactures these (and guarantees each tube for three years), also makes lamp bulbs that are guaranteed for one year and are as bright as the others. We have some that have lasted at least two years.

We feel they are really trying to manufacture a quality product and go against the trend toward making things that don't last. To get these lamps, contact Gerry Cowan, 1320 Zachary Road, Roslyn, Pennsylvania 19001. (215) 886-3270 or home (215) 657-2336. He will send information. Please mention ARF to him, and he will also help us with a donation.

SPIRIT DREAM:

Over the months, I have come to know Estelle Burroughs, a clairvoyant with gifts of healing. When she learned that the famous British medium Edna Cox was coming from England to visit her, she lovingly arranged for me to meet with Edna. I looked forward to that meeting for some weeks, hoping to really get some direct and definite leadings from her which would help the project and also be evidential and convincing.

The occasion was an immediate disappointment (and I told her so). She could give me nothing provable, and only told me to try to develop my own psychic abilities by learning to listen to what loving and intelligent spirits were trying to tell me: that I must make an appointment with those spirits at the same time each day when I would meditate and wait for something to come from them. I had read that in books, but it seemed impractical with my irregular schedule. But Edna insisted, and finally I remember that I often wake up at about 6:30 each day and go back to sleep again. I agreed with Edna that I would ask the spirits to wake me at 6:00 A.M. each day, and that I would spend time in meditation before going back to sleep.

The next morning I awoke (no alarm) precisely at 6:00. Nothing came. Second morning the same thing happened, and the 3rd morning (May 15, 1970) I also woke within a minute of 6:00 and soon went back to sleep.

Then I had a strange dream. In my dream, I first of all found myself in a community on land which was in some way similar to or connected with the Hutterites (who lived in 200 farming communes in Canada and the west). I then found that people there were very interested in the work I am trying to do. (They also gave me very good food to eat.) The strange part of it was that this community was less than a day's drive from Philadelphia, and having known of the Hutterites for 22 years, I was certain that they had nothing in this area.

In the second part of the dream, I found myself in the Forest River Hutterite Colony in Fordville, North Dakota, where I had not been since 1955 or 1956. There (in the dream) I was told that Joe Maendel (with whom I had traveled across the country at that time) was now quite ill in a strange way, in that he was (or was believed to be) in a terrible state on confusion, and that the community in which he was did not know what to do about it. I was only **told** this (in the dream) and was not able to speak with Joe to find out if it were actually so.

I thought about the whole dream when I awoke and felt it to be significant, but so very difficult to verify. That a Hutterite group of any kind was near here was an obvious error in the dream, and if I wrote to Forest River, they would probably not answer, especially if the confusion was real. I couldn't afford to telephone, so I did nothing.

Next morning (and the following days), I did not wake at 6:00, though I had asked to be woken. Then about three days later, I received a letter from Kathe Quick, who teaches English at the University of Iowa. She had read our pamphlet and had written to me once or twice before. She had lost the pamphlet and the letter was a bit confused, but there was one paragraph that struck my eye at once. I hope she will pardon me if I quote her letter (written May 15, mailed May 16).

"We are still undecided about next year. If we teach at an alternative school, it will most likely be Buck Brook in Roscoe, New York, a new branch of George Von Hilsheimer's Orange City, Florida school for troubled children. His group is a sort of Hutterite offshoot and they work a pretty structured scene. (No smoking of **anything**, no freaky clothes, strictly for good public relations, etc.), but the results with the kids are fantastic. We'd get room and board and $7.50 a week."

The word "Hutterite" at this moment amazed me, and I telephoned Edna Cox to tell her of the dream. She said she had been praying for me because the evidence I needed had to be in my own experience. She asked me if I had prayed for Joe Maendel when I woke from the dream, and I admitted that I had not. She admonished me severely that I should pray for him, and told me that the spirits could not help me further if I did not properly use what they had already shown me. I began to pray for him, and it became clear to me that I should phone Forest River that night and find out if Joe was there and really in a state of confusion. I did that and, to my amazement, his situation was said to be exactly as I had dreamed it! It was almost as if they had expected me to call, and they quickly brought to the phone a member who had known me when I first met the Hutterites in Canada way back in 1948. His name was Josua Hofer, and he remembered me well and asked that I write to him, he then told me that he was in Forest River because of the situation with Joe, that the community was split and they did not know what to do about it.

Of course he asked me how I knew, and I told him that I had dreamed it, which he absolutely would not believe (but the one thing he remembered about me was my openness). He thought I must have heard about Joe's situation from his daughter, but I haven't seen them in years and have no idea where they are.

Josua asked me what I was doing and I said I would write to him and send him a pamphlet, but that he might not like it; that I would also write to Joe. But they asked me not to write anything which would confuse him more. All I can do, however, is tell the truth (as I experience it) and pray that God will show him the truth (and all the others also.) I have always felt that the Hutterites must have a great and important task in this country, and that their 200 or so communities are not here without a purpose. I ask that they will all be led to something great and new.

When I told Josua about Edna Cox, he seemed fearful of that, saying that in Old Testament times they killed such people, and that he believed they were right. But it seems to me that he fails to distinguish between people involved in magic (for selfish reasons) and those who seek and receive insights from good spirits for loving and unselfish purposes. Certainly the dream that came to me was not the result of any kind of magic.

Of course, I had to call Edna again and tell her about the phone call, and she admonished me to continue praying for Joe, which I do. The next morning I did not wake at 6:00 either, but I had it said to me in a dream or while waking up that in about two weeks time the house would be full of people who would be here because of the project. So I marked that on the calendar and am trying to prepare for it.

Since the other part of the dream and the letter from Kathe Quick seemed to confirm one another (I find that Buck Brook is only 160 miles from here), it seemed clear that I should follow through and go visit Buck Brook School, and I will phone them tonight and see if that is possible.

Other leadings have also been given to me in recent times, and I have been advised to try to teach other people some of the things I am discovering, either by setting up a school here or getting some classes going somewhere else, and that seems to be the first thing to do now. It's hard to start a class in "breaking down resistance to change" as people would not understand that, and I don't really know how to do it myself yet. So I have been trying to start a class in "psychic research," or, the scientific study of psychic phenomena. In advertising the course, it is clearly said that we will not be involved with magic, but only in finding out how these insights can be used in a loving way to prevent a civil war and help the New Age of peace to come quickly instead.

NEW HOPE FOR THE DEAD:

By far the most difficult and, possibly, the most important project is to reestablish the work of Carl Wickland, M.D., who wrote **30 Years Among the Dead**, a book based on his many years of medical practice as one of our early psychiatrists. Dr. Wickland worked very closely with an excellent medium who happened to be his wife. Together, over a long period of time, and in the presence of many witnesses, he was able to show that many "psychosomatic" illnesses and some cases of "insanity" were actually the result of people being obsessed by living spirits of humans who have lost their physical bodies (or died).

Dr. Wickland, who died in 1945, has done the most evidential work I have yet found which clarifies the nature and reality of the spirit world. His work shows how ignorant spirits can cause all kinds of confusion and suffering to people here, and that there are several ways in which they can be removed from the person possessed (prayer is one way), and that this can help to free the ignorant spirit to find the spirit world as well as cure the illness of the affected mortal. It seems evident that Wickland's work fully confirms the New Testament reports in which Jesus and the Apostles were able to cure some illnesses by those means.

THE "HUTTERITE DREAM"

By following Edna Cox's instructions, I was given a dream in which I saw myself in two communes; the first would be less than a day's drive from here, and in some way connected with the Hutterites, the oldest communal group in America. In the dream, my visit there seemed to be very important.

The second commune was a Hutterite colony in which I had been 14 years ago. I dreamed I was told that a friend there was in a terrible state of confusion and that the community was very troubled by his situation. It all seemed fantastic, so I did nothing about it until three days later, when a letter came from Iowa confirming the existence and giving me the location of the first group. That led me to phone North Dakota, and the phone call amazed me by exactly confirming the second part of the dream. This, in turn, led me to visit Buck Brook Farm near Jeffersonville, New York. There again many things happened just as in the dream, and, in addition, I met one girl (Laurie) who turned out to be a psychic, able to see the haze around the poles of a magnet, as well as have other experiences. She had been considered schizophrenic since she was 12 years old and spent five years in an institution. She had never before heard that her psychic experiences might be good and not a sign of insanity. The staff at Buck Brook knew little of such things, but they were open. They always told her she was not "crazy," and they are convinced that love and freedom can cure many ills. The tremendous success they have with "emotionally disturbed" students shows that their methods work. They have about 45 students and 10 staff members. A later visit to Buck Brook with Maria Youngblood

showed that Laurie was able to see auras around people and plants.

BEHIND THE IRON CURTAIN; PSYCHIC DISCOVERIES:

The publication of the book, **Psychic Discoveries Behind the Iron Curtain** (Ostrander and Schroeder) by Prentice-Hall has opened new fields. In Russia, Bulgaria, Czechoslovakia, the government finances psychic research and has some of their best men working at it full time. They have photographed auras around plants and parts of the human body, and I have found it possible (with Laurie at Buck Brook) to prove that they are telling the truth. Experiments showed that when part of a leaf is cut off, the aura of the whole leaf is still visible. Laurie was able to describe that as something she can see before I showed her that Russian scientists have proved it photographically. A living organism seems to have what is described as "the biophysical plasma body" in addition to the physical body; it appears to be the same things that psychics have spoken of for years as a "spirit body." There are other far-reaching, carefully investigated psychic discoveries in the socialist countries that have been of some special importance; even machines that are apparently energized by forces from the human mind. One of the most far-reaching discoveries seems to be:

Astrological birth control, which is the work of a Czech psychiatrist, Eugen Jonas. At first they thought he was crazy, but according to the book, his work is proving successful and is now financed by the government. Not only is he able to prevent unwanted pregnancies in a natural way, but he has also found ways to eliminate birth defects, retardation and miscarriages by determining the best time for a woman to conceive according to the specific time of her birth.

I do not believe in astrology, but I have to admit that if these claims are proven true, I will have to reconsider my beliefs. If astrological birth control (ABC) works, it should have a tremendous effect on this country because:

1) Many people will be freer to act upon the real beliefs and insights that they have and a lot of fear will be gone. Truth will emerge.

Therefore, we decided that our first responsibility in this question was to get the facts which were not included in the book. Maria and I visited Alan Vaughn, Associate Editor of **Psychic** in New York City, and he gave us the address of the authors of the book. We phoned them and were told that a study of 22,000 women was now completed, and that ABC was shown to be about 100% effective. But they said the books telling how it is done were not yet translated. Still? Two years after they got them? So they say.

Also, we are in contact with Prentice-Hall's editor, who tells us they are negotiating for a book from the girls who wrote **Psychic Discoveries Behind the Iron Curtain** in which the "nitty-gritty" of ABC will be explained, so they also seem to be taking it seriously.

MORE HOPE FOR THE DEAD:

Carl Wickland's method of proving the reality of the spirit world is something I would like to re-establish and record on TV tape, but it has become clear to me that that is not the first task of the moment. I have just received a list of six people who are probably mediums that might be interested in what we are doing. If any of them come, that'll be a clue, as a good medium would be required to make that kind of work possible, I believe. **Thirty Years Among the Dead** by Dr. Wickland is well worth reading.

The American Book of the Dead is a title of a small book that I feel ought to be published as soon as possible. It will describe in plain language all we know of what seems to happen to people when they die. Far more is in fact known about that question than most of us may realize. I hope to write, print, and publish it myself.

A VISIT WITH W.D. CHESNEY IN WISCONSIN

In the book **30 Years Among the Dead**, Dr. Carol Wickland revealed that spirits sometimes influence people who are thought to be "mentally ill," and he was able to show that it was possible to cause the obsessing spirit to come to the forefront and speak out, give its name and place of former life, the time of death, and a great deal of additional evidence which could be verified by ordinary methods. Dr. Wickland died in 1945, and in order to confirm his work and obtain more information about it, I gave some effort to locate people still living who knew more of the details than I was able to find in Dr. Wickland's two books. (**Gateway of Understanding** was his last book, published in 1934, but now scarce.)

In searching for people who knew the work or were present at the Wicklands, I first contacted a Rev. Titus in Rochester, New York, and then came into correspondence with Dr. Chesney in Wisconsin. Rev. Titus fully confirmed the validity of Wickland's work, but he was only an occasional visitor, so he was not able to give me the scientific facts which were required. Dr. Chesney, on the other hand, was already something of a scientist and an inventor (He was probably the first man to synthesize vitamin D, one of the developers of infrared photography, and a personal friend of Steinmetz and Edison. He is now over ninety.)

Dr. Chesney was able to tell me some of the exact details of the way in which Dr. Wickland used static electricity to cause obsessing spirits to make themselves evident, and to want to leave the person they were obsessing. Chesney himself learned a lot about the reality of the spirit world from his father, Judge E.E. Chesney, who was a close friend of Abraham Lincoln. There is considerable historical evidence that Lincoln also knew quite a bit about these matters.

While with Dr. Chesney, I was able to make several hours of tape recordings of the things he was able to tell me, and I left the tape recorder with him so that he can record more material to send to me at a later time. I also came to see the possibility of doing further scientific experiments to confirm some of the discoveries which Dr. Chesney and Dr. Wickland were able to tell about.

WHALES: Hitchhiking on the way back from Wisconsin, I was picked up by a boy and girl from the University of Wisconsin at Madison who were driving to Connecticut in an old station wagon, and we shared the trip together (I helped with driving and mechanical problems) all the way to Eastern Pennsylvania. They had with them a few copies of **The Saturday**

Review, and I took the time to read an article by Paul Ehrlich entitled, "The Myth of Food From the Sea." In it he described the way in which whales are becoming an extinct species because they are being overexploited. It seems that the governments of whaling countries are well aware of the problem, and a commission was appointed to study it. Biologists were employed to make careful studies of the number of whales and whether they could reproduce at the rate they were being killed, and it was shown that they could not. Predictions were made that the catch would diminish, and lower quotas were suggested by the scientists. The predictions of the scientists turned out to be amazingly accurate, and it was seen that unless the number of whales taken was markedly reduced, they would soon become extinct. Nevertheless, after the scientific facts were fully established, the commission set the quotas twice as high as the scientists had suggested (or about the same as before) with the result that the actual catch is declining and whales are indeed headed for extinction.

I have to admit that whales are not my chief concern, but I was deeply shocked by the implications of these findings. As I waited, (sometimes hours for a ride) I was able to see the horror of the spectacle that confronts us. Logic would have it that even for their own interest alone, the whalers would want to save the whales, but instead each of them says to himself, "If I don't take him, then the other fellow probably will, so I might just as well take them while I can."

EQUIPMENT NEEDED—In order to do some of these experiments, we will need some equipment to produce very high voltage electricity such as might in earlier times have been produced by a static electric generator, but which might now be produced by more sophisticated methods. In those days the high voltage could be produced with a Wimhurst machine or a Toepler-Holz plate machine, and either of those may still be satisfactory if they could be obtained today.

However, it is quite likely that either of those machines might produce considerable interference with the neighbors' television reception, so it would be far better if some more modern type of high voltage equipment could be obtained. Dr. Chesney spoke of equipment having a maximum potential of 800,000 volts, DC, but such a low current that nothing more than a severe shock would be given to a person who came close to it. However, I believe that so much voltage is not actually required, and the maximum length of spark which was produced was never more than ten or twelve inches. It may be that the kind of high voltage power supply that is used in some projection television equipment would be adequate.

ANOTHER PIECE of equipment which is needed for work which will be described below is a camera with lights designed for infrared photography. For that any ordinary camera will do which can take infra-red film (would probably have to be a 35mm) and it would probably have to have a large lens opening. A half-frame 35 would probably be the very best because of its greater depth of focus at close range. In addition, an infra-red floodlight system suitable for use with such a camera would be required. The light would only have to be sufficient to take pictures at a distance of twenty feet at a speed of about 1/50 of a second.

PHILADELPHIA BULLETIN REPORTERS & A SEANCE AT SILVER BELLE

At one of our last classes in psychic research (which ARF conducts at a local church), we had a visit from a reporter and a photographer from the Philadelphia **Evening Bulletin**, and they turned out to be quite friendly. Helen Bauhof, the photographer, was also exceptionally interested in the subjects we were talking about, and seemed to have quite a bit of knowledge in the field—also some personal experience which she freely reported to the group. She told of having been sent to take pictures of civic functions for the paper and found when the film was developed that one of the people in the picture who definitely was present when it was taken was not there in the negative. She was terribly embarrassed about this as the pictures could not be repeated, and the missing person was the main subject. A few days later, she says, that person suddenly died.

I told Helen that I would like to see the negative and have prints of the pictures made so that I could take them to others in the picture and find out if the facts could be verified. She has promised to find the negatives for me as soon as she is able to.

At the class Helen also told us of some spiritualist seances which she had attended, and mentioned a camp or hotel called Silver Belle in Ephrata, Pennsylvania, where GOOD mediums could be found. She said that I would have to get there quickly because it was the end of the season and the camp would close at Labor Day. As it was only 65 miles away, I went out there at the first opportunity and found that Helen's information was correct.

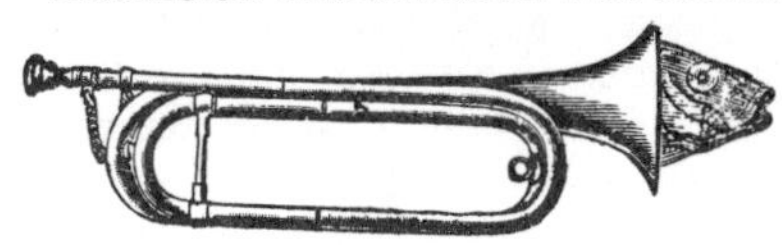

MEDIUMS

At Silver Belle, mediums and spiritualists of different kinds come together each summer from all over the country. In a short time I was able to speak to a young man named Reed Brown, who seemed to be by reputation the best medium at the place. We talked for about ten minutes and I was able to tell him about the work I was doing. He was not unsympathetic, but said that he was entirely booked up for two years to come, and could not be of help. He suggested that I go to see Roy Burkhardt, which I was then able to do. It took Roy a little while to understand the relationship that I saw between the present sociopolitical situation and the world of spirit realities. He told me that he could do nothing at that time in terms of establishing contact with spirits that could help me because it was the end of the season, and the constant work of trance mediumship had left him too exhausted to take on any new work at this time. However, he assured me that he was interested in what I was doing and would find ways in which he could help me. He told me that he lives

in Kansas City, Missouri, where many young people have come to take a great deal of interest in his work, and that he would tell them about what I am doing and give them copies of my pamphlets and other things I might be able to send him. Possibly some of them would come and work with us full time.

During the afternoon in Silver Belle I attended two chapel services. The first was very interesting in that the speaker was able to give specific messages to specific people in the audience, and it seemed obvious that it was not a trick or a plant. The people knew the names of the spirits (loved ones they had known) that he told of, and he was almost always correct. I hoped that the speaker would get something for me, but he did not.

The second such chapel service was much less evidential, and the psychic was not able to give any names at all, but some of the material he got seemed correct. He did seem to get a message for me, but it seemed entirely irrelevant and vague. He claimed that he saw the spirit of an old minister with a Bible telling him that I would get the "position" that I was seeking. He claimed that I knew this minister as a child. I told him that it was extremely unlikely, my being Jewish.

After that, I happened to meet Israel Schotz, who had been something of a medium in the past, and he told me about the "trumpet seances" which took place there and suggested that I attend one (which I did). The seance was held by an older man who had recently been quite ill and was only there on a visit. The group was limited to about ten people, each of whom payed $5 to attend. The medium explained that the seance would be carried on as follows:

He would be awake, not in trance. The people would each fill out on a little business card the names of loved ones they wished to contact, and that person's relationship to them. Then a single question could be written in the hope that the spirit would answer it vocally or in writing on the card. During the seance, the room would be totally darkened and the spirits would carry around a sheet metal trumpet (a cone about 3′ long, 1″ at small end, 6″ at large end, and made in three sections which could be separated). They would, he expected, in most cases be able to place each person's card (which was only the size of a business card) inside the large end of the trumpet, and it would in that way be carried directly to the person to whom it belonged. While doing that, the spirit would tap on the end of the trumpet so that it would be possible for the sitter to locate the trumpet in the dark and take his card. Of course it would also be possible to ask questions of the spirit and receive vocal answers.

The seance then began with all of us reciting the Lord's Prayer and singing a hymn, after which it soon began to happen that we heard the voices of various spirits that different people had asked to hear from. Between the time that we wrote and signed the cards and the darkening of the room, the medium talked to us for quite a while and said that he was doing that to pass the time so that the spirits that were called upon would have time to get there. Before he turned out the lights, the medium shuffled the cards, but it was apparent that he could not have read any of them without some very special tricks, which did not seem possible while he talked. There was no part of the room which was not visible to all of us. There was also no way in which the medium could, as far as I could tell, have read the cards in the total darkness of the seance.

As the seance progressed, it did seem that many spirits came and spoke to various of the sitters; yes, in fact, each one was contacted and greeted by the person whose name they had given, and each was given a card which turned out later to be the correct card, and each card did also bear a written message, supposedly written by the spirit. Most of the people—all except myself actually—only asked questions concerning the welfare of their loved ones on the other side, and they were all satisfied with their answers, and accepted the veracity of the experience. I had met three of the group some time before and talked with them, and had the feeling that they were honest, or, at least not part of any trickery. My own case was different. I wanted not only to get some evidential proof of the veracity of the spirit appearances (one spirit did actually seem to appear) but I also wanted some information as to how I could find the help I needed to carry on my work. Before the seance began, I was warned by the medium that this was not to be an experimental seance, as he sensed that I was looking for that by something I asked.

I had asked for contact with three spirits, namely, Dr. Carl A. Wickland, my uncle, Mike Silvert, and one Yogananda, whom a clairvoyant that I trust had told me she saw at my side, and who might be a guide to me. When it came my turn things did not go too well. Having been asked to give the full name of our relative, I had given my uncle's name as Meyer Silversweig, which is what he had originally been called by his parents. But to us he was always just Mike, not even Uncle Mike. He died a suicide in the 1950's. Before that he had been a psychiatrist and the right hand man of Wilhelm Reich. When Reich was imprisoned because he refused on the grounds of conscience to obey a court injunction to stop publishing his discoveries about Orgone Energy, he was sentenced to a year in prison for contempt of court. Mike took the same stand. Reich died in prison, but Mike was released, though no longer able to practice medicine or do the work he had been doing before.

When my turn came, I was greeted by a voice which simply said, "Hello, Art, I'm your uncle Meyer Silverns" or something like that. The name was definitely not "Silversweig" as I had written on the card, but it was also not quite "Silvert" as I would have expected. Instead of accepting him with the love and joy I probably should have, I asked him to repeat the name, and that caused the medium (who was not in trance) to be quite angry. All through the seance, I had the definite impression that the medium always spoke from the table where he was, and often did so immediately after a spirit had spoken from another part of the room (or so it sounded). This time it was the same. After that there seemed to be no further word from my uncle (if it were he). Then the medium said that someone was there for me who said his name was Toganananda or something, and asked if I knew who he was. I said that possibly he was a guide to me, and the medium seemed upset that I was not sure who my guides were, or seemed not to believe that I was being honest, and said I was a troublemaker. Then a voice came through which identified itself as Dr. Carl Wickland, and he said he was my guide. I asked him what he thought of my plan to go to Wisconsin to see his old friend, and if he knew who I meant (assuming that if he was my guide, and I was going to see Dr. Chesney), he would know that and indicate the same, but instead he seemed not to know whom I was going to see, so that also seemed inconclusive. Later, though, he came through and told me that he would like to shake hands with me because he appreciated what I was doing and I was able to reach into the mouth of the trumpet, and shake what felt like a warm human hand of a man (which could theoretically have been that of the medium—if he could have been near me, and the desk almost simultaneously).

One of the spirits then brought me my card in the trumpet, but I dropped it. The medium said "You would."

We found the card when the lights came on, and it was signed —supposedly by my uncle with the name, "Uncle Meyer," which seemed indicative of a gross inaccuracy. I had the nerve to mention that point to the group, which made the medium furious, and he actually pushed me out of the room—though he at first offered to refund my money (which I did not ask).

On later consideration however, I have to admit that it is perfectly possible that everything was on the level, since Mike has been a spirit now for many years, and, as far as I know, since that time no one has gone to the other side to whom he would have been known as "Mike." If he has come to know his parents again, they would surely be calling him "Meyer," and not having heard the name "Silvert" for so many years, it would be reasonable to assume that he had become a little uncertain of what it was like. In Wickland's book, there is a lot of evidence that many spirits do forget a great deal of the things that we would remember about them. So, I really have to see that I might also have been at fault in the matter. Nevertheless, it was not necessary for the medium to ask someone to leave and even push them out because they had doubts, and Israel Schotz could hardly believe that he had done that.

PUPPIES PUPPIES PUPPIES

A great donation came from an enchanted princess who lives here as a dog called "Lady." She has presented us with 13 puppies, which are full German shepherds (but she does not believe in papers). One was weak and died before even being able to suck, and another was sort of thin, and might have survived if we had given it enough special attention, but I did not know of the need until it was too late. The remaining eleven are fat and happy and will have their eyes open in a few days.

STARTING A SCHOOL

The result of the Buck Brook Visits has given me the idea of starting a free school similar to Buck Brook (no classes), but financed in other ways, and having as its object to let people who are sick of high schools or too clever for colleges study about the future instead of the past. How about a school whose students do independent research on finding out how to prevent the next war instead of how the previous one was fought? I have already written to the New Schools Exchange Newsletter (Pettigrew, Arkansas 72752) about this and will soon be writing about it to the Teacher Drop Out Center (Box 521, Amherst, Massachusetts 01002) to ask for teachers who would like to come and help start such a school.

Already I have had a visit from Tony Pearce of Friends World College, and they are thinking of sending some of their students here in October to work with the project as part of their work-study program.

Another free school idea which came to us recently is reminiscent of the old German apprenticeship idea. It is called THE APPRENTICESHIP PROGRAM, and is to publish a newsletter for high school students, and those who want them. The student would list what they want to learn and would contact someone who can teach it, and then, with parent's permission, they would go to live with that person or family and work together with them, learning the work there. The work might be organic farming, building, or some profession. Room and board would be provided by the family who extends the invitation in exchange for help of the student as an apprentice; other expenses would be paid by the parents. Progress reports would be sent by student and teacher to a school in California in which the student would be enrolled as doing independent study, and that school would issue the high school diploma. In that way the whole thing would be legal. The address of the A.P. is 12100 Skyline Blvd., Los Gatos, California 95030 Tel. (408) 867-2260

THE MOON seems to have a lot on. The enclosed article was reprinted in part from **Argosy**, a magazine that has, from time to time, published articles which are found nowhere else and may be of interest to our readers. The article gives a chronological catalog of lunar events, which we got and copied a page for you. To get one, send $3 to Clearinghouse for Federal, Scientific, and Technical Information, Springfield, Virginia 22151. Released by NASA.

"MYSTERIOUS MONUMENTS ON THE MOON"
by Ivan Sanderson

Four years ago, Russia's Luna-9 and America's Orbiter-2 both photographed groups of solid structures at two widely separated locations on the lunar surface. These two groups of objects are arranged in definite geographic patterns and appear to have been placed there by intelligent beings. Since American space officials have chosen not to publicize these findings, our readers are probably not aware of their existence.

The Luna-9 photographs, taken on February 4, 1966, after the craft had landed in the Ocean of Storms, reveal two straight lines of equidistant stones that look like the markers along an airport runway. These circular stones are all identical, and are positioned at an angle that produces a strong reflection from the sun, which would render them visible to descending aircraft.

Upon examining the photographs, Russian scientist Dr. S. Ivanov, winner of a Laureate of the State Prize (equivalent to a Nobel Laureate), and inventor of stereo movies in the U.S.S.R., noted that a chance displacement of Luna-9 on its horizontal axis had caused the stones to be taken at slightly different angles. This double set of photographs allowed him to produce a three-dimensional steroscopic view of the lunar "runway."

Why the Luna-9 station changed its position between its second and third transmissions is not known. The official Russian explanation was: "Deformation of the lunar surface. The ground may have settled at the spot where the station landed, or perhaps a small stone caused the initial instability."

Whatever the reason, it was good luck for the Russian observers. "With the stereoscopic effect," reported Sr. S. Ivanov and Engineer Dr. A. Bruenko, "we can affirm that the distance between stones one, three, two and four is equal. The stones are identical in measurement. There does not seem to be any height or elevation nearby from which the stones could have been rolled and scattered into this geometric form. The objects as seen in three-D seem to be arranged according to definite geometric laws."

The second set of photographs were taken by America's Orbiter-2 on November 20, 1966, twenty-nine miles above the lunar surface, over the Sea of Tranquility. The photographs, of an area some 2,000 miles from the "runway" reported by the Russians on the Ocean of Storms, show what appear to be the shadows of eight pointed spires shaped like Cleopatra's Needle (the ancient Egyptian obelisk now in Central Park in New York) and the Washington Monument.

Since Orbiter-2's cameras were pointed straight down at these spires, only their shadows are visible; but NASA stated that the sun was eleven degrees above the horizon, and from this, American space scientists estimated that the "largest protuberance" is approximately fifty feet wide at the base and from forty to seventy-five feet high.

However, the Russian scientists who examined the Orbiter-2 photos disagreed with these American estimates, and said that the smallest of these eight apparent obelisks was "similar to an extremely large fir tree," while the largest was, by their estimation, three times the height claimed by the Americans—or as tall as a fifteen-story building!

In addition, Soviet Space Engineer Alexander Abramov has come up with a rather startling geometrical analysis of the arrangement of these objects. By calculating the angles at which they appear to be set, he asserts that they constitute an "Egyptian triangle" on the moon—a precise geometric configuration known in ancient Egypt as an "abaka." "The distribution of these lunar objects" states Abramov, "is similar to the plan of the Egyptian pyramids constructed by Pharaoh Cheops, Chephren, and Menkaura at Gizeh, near Cairo. The centers of the sphires in this lunar 'abaka' are arranged in precisely the same way as the apices of the three great pyramids."

What is America's position regarding the investigation of the mysterious moon objects? A high NASA authority, when questioned on what has been done by us in the four years since these objects were photographed, replied, "Yes, we know of these photographs and they were very clear, but there has been no speculation on them, and they have been filed for now."

We fail to appreciate such an attitude toward something of the first interest to our whole moon-probing endeavor; especially one which has been so openly investigated by our rivals in this effort.

However, in 1968, NASA released a very remarkable publication, the "Chronological Catalogue of Reported Lunar Events." This list includes lights, both stationary and moving, appearing on the moon and then suddenly disappearing; some perfectly circular craters that look more like domes and which are, in some cases, arranged in perfect alignment, and such phenomena as glowing mists, and sudden patches of gem-colored outpourings.

John O'Neil, former science editor of the late New York **Herald Tribune** said that he observed a gigantic bridgelike structure in the Sea of Crises (on the moon) under which the sun shone when at a low angle. This was later confirmed by other astronomers. Recent literature has been crammed with descriptions of wall-like structures that form perfect squares or rhombs; of rills that look like water-eroded river beds, and of lines that look like roads, and even a grid of streets on the lunar surface.

Of course, many such apparently man-made structures, seen on earth from on high, later prove to be simply natural formations. At the southern edge of the Sahara Desert in Northern Nigeria one such natural formation, when seen from above, appears to be an enormous layout of structured walls. And from an orbiting space craft, the Barringer Crater in Arizona looks like an artificial construction.

Despite both manned and unmanned landings on the moon, we still know very little of just what is on its surface, simply because of its size, the wide variety of its topography, and the still comparatively limited range of our probes.

Suppose for a moment that the earth was probed from the moon by two manned landings—let us say in Utah and East Africa. How much would we learn of the surface of the rest of the planet? Our oceans, lakes, mountain ranges and rivers might well be spotted and photographed from on high on approach, but just how many smaller things like pyramids and obelisks, built by thinking creatures with an intelligent design in mind, might go unrecognized; or, still worse, might even be seen and filed away simply because they did not accord with what was accepted as possible by the experts at the home base?

Though very little attention has been given to the mysterious moon objects in this country, both the Orbiter-2 obelisk photographs and the Luna-9 runway pictures were widely published in the Soviet Union, for Russian scientists have always been extremely interested in the pursuit of any evidence for extraterrestrial life. Moreover, the question the Russian scientists are now asking about these lunar objects is whether intelligent beings could have visited our moon long ago, and erected thereupon permanent monuments and landing fields.

This question should not come as any great surprise to us because the Russians have long and consistently gone after archaeological and historical evidence of superior life forms having visited this planet.

The material amassed by them in support of such an idea is now somewhat overwhelming. For instance, they claim that many biblical stories, such as the apparent destruction of Sodom and Gomorrah by an atomic bomb, are historical accounts of such visitations. They have reported finding metallic discs, like modern recording platters, in Asiatic caves, and they have published reproductions of early Christian wall paintings from old monasteries in Yugoslavia that appear to show angels in space ships.

There are quite a number of strange material things that strongly suggest some extraterrestrial origin or influence. The Egyptian pyramids continue to puzzle scholars, who are now

Sample Page

CHRONOLOGICAL CATALOG OF REPORTED LUNAR OBJECTS

No.	Date and Time	Feature or Location: Duration	Description	Obeserver	Reference
528	1965 Oct 13, 3 hr. 2 min.	Aristarchus	Pale, blue-violet tint on E wall bright spot and whole length of E wall; pale violet radiance in crater, exclusive of 3 region. Dark violet, nimbus.	Bartlett	Bartlett 1967
529	1965 Nov. 15	Aristarchus	Bright spots	L. Johnson	**Phys. Today** 1966
530	1965 Dec. 1		Reddish glow followed by black obscuration.	Evrard and others	Gingerich 1966
531	1965 Dec. 4, 4 hr. 25 min.	Ross D	Obscuration of part of rim, also bright area 7-10 km diameter, not seen on following night (4 hr.-7 hr. 30 min.)	Cross (Harris, Cragg on Dec. 5)	Harris 1967
532	1966 Feb. 7, 1 hr. 10 min.	Aristarchus	Nimbus only; intense violet hue.	Bartlett	Bartlett 1967
533	1966 Mar. 29, 21 hr.	Archimedes	Floor bands brilliant	E.G. Hill	**B.A.A. Lunar Sec. Circ.** 1966, **1**, No. 6
534	1966 Apr. 2, 23 hr. 30 min.	Aristarchus 20 min.	Central Peak very bright.	M. Brown	**B.A.A. Lunar Sec. Circ.** 1966, **1**, No. 7
535	1966 Apr. 3, 23 hr.	Aristarchus 30 min.	Central Peak very bright.	M. Brown	**B.A.A. Lunar Sec. Circ.** 1966, **1**, No. 7
536	1966 Apr. 12, 1 hr. 5 min.	Gassendi 18 min.	Abrupt flash of red settling immediately to point of red haze near NW wall. Continuous until 1 hr. 23 min.	Whippey	**B.A.A. Lunar Sec. Circ.** 1967, **2**, No. 5
537	1966 Apr. 30-May 2	Gassendi	Red glows.	Sartory, Moore, Moseley, Ringsdore	**J.B.A.A.** 1966 **B.A.A. Lunar Sec. Circ.** 1966, **1**, No. 6
538	1966 May 1, 21 hr. 55 min. 22 hr. 45 min.	Aristarchus 50 min.	Red patch.	Patterson	**B.A.A. Lunar Sec. Circ.** 1966, **1**, No. 6
539	1966 May 1, 22 hr. 10 min.	Aristarchus 15. min.	Small intense white spot NW of crater wall.	M. Brown, Sartory	**B.A.A. Lunar Sec. Circ.** 1966, **1**, Nos. 6, 7
540	1966 May 27, 21 hr. 10 min.	Alphonsus 50 min.	Faint red patches.	Sartory, Moore, Moseley	**B.A.A. Lunar Sec. Circ.** 1966, **1**, No. 6

studying their sealed rooms with the most advanced and sensitive electromagnetic devices. One scientist, Dr. Amr Gohed, stated officially to the London **Times** that "Either the geometry of the pyramids is in substantial error, which would affect our readings, or there is a mystery which is beyond explanation . . . there is some force that defies the laws of science at work in the pyramids."

The possibility of extraterrestrial influence on the moon was put forth at a meeting of the American Rocketry Society by Dr. Carl Sagan, who said, rather simply, that "Intelligent beings from elsewhere in the universe may have—or have had—bases on the averted side of the moon."

Why, we may ask, did man ever start making obeliskes anyway? It's a very tough job and seemingly purposeless. Is the origin of the obelisks on this earth, and those on the moon, the same? Could both be ancient markers originally erected by alien space travelers for guidance of later arrivals?

The basic information contained in this article was given to us by Lynn Schroeder and Sheila Ostrander, who last year visited Russia, Bulgaria, Romania and Czechoslovakia, at the invitation of Russian scientific organizations, which had set up an international symposium covering the general field of paraphysics. They are the joint authors of a book on this subject, published this month by Prentice-Hall, entitled **Psychic Discoveries Behind the Iron Curtain**. The substance of this article is extraneous to this fascinating book, but it seems to us to be of great interest in this day and age of moon and planetary probes by the USSR and ourselves.

While in California recently, I met a scientist who had been employed by NASA several years ago, and he had helped to design the Houston Space Center. He told me that at about the time of photographing the shadows of the obelisks discussed in this article, the Boeing Aircraft Corporation published the same photo in their company newsletter. He had not seen it published since then. He also told me that while working for NASA, he found it uncommonly difficult to get information from them; even such information as was required for the intelligent design of the center. Unfortunately, I do not have permission to publish that man's name here, but I do have the impression that he was a reliable person.

In the body of the article, Ivan Sanderson mentions a document called "The Chronological Catalog of Reported Lunar Events" and tells us some of the things found there. Below is one page of that catalog (which we have). It has been "released" to the public—but as far as I know, it has never been printed.

If monuments have been discovered on the moon, one would suppose that this is about the most important result of the whole moon probe effort. Why is this information not widely published by NASA? Why are they not fully investigated (or are they)? What other type of information is being withheld? Why? Should not NASA be investigated? Also, why did the Russians also hold back on these discoveries for four years?

NEW PUBLICATION

A good book just out: **BIOFEEDBACK** by Karlin and Andrews (published by Lippincott).

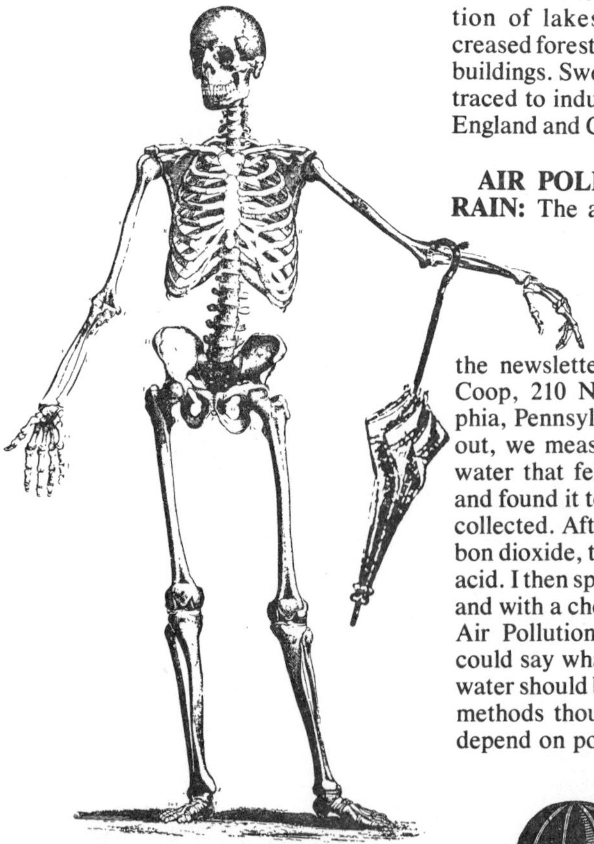

ACID RAINWATER Rain in the northeastern United States has become disturbingly acid, according to a report in the **New York Times**. The **Times** story, written by David Bird and based on information supplied by three New England college professors, says tnat rain samples from a seven state area were 10 to 100 times more acid than expected. The states are Maine, New Hampshire, Vermont, Massachusetts, Rhode Island Connecticut and New York.

The researchers, Dr. Gene E Likens of Cornell, Dr. F. Herbert Borman of Yale and Dr. Noye M. Johnson of Dartmouth, said that rain samples showed pH values of between three and five, both strongly acid reading on the scale used to measure acidity or alkalinity.

According to the three scientists, the acidity can be attributed to certain air pollutants, namely sulfur and nitrogen oxides. These compounds can be converted chemically to strong acids in the atmosphere.

The professors observed that although there is no apparent threat to human health from acid rainfall, it can do considerable damage to man-made structues, soil and ecosystems. Studies in Sweden have already confirmed the link between acid rain and loss of nutrients from soil, acidification of lakes, salmon deaths, decreased forest growth and corrosion of buildings. Sweden's acid rain has been traced to industrial air pollution from England and Germany's Ruhr Valley.

AIR POLLUTION AND ACID RAIN: The above is reprinted from the newsletter of the Ecology Food Coop, 210 N. 36th Street, Philadelphia, Pennsylvania 19104. To check it out, we measured the acidity of rain water that fell in West Philadelphia, and found it to have a pH of 3.5 when collected. After boiling to remove carbon dioxide, the pH was 4.5; still quite acid. I then spoke with a meteorologist and with a chemist at the Philadelphia Air Pollution Laboratory, but none could say what the pH of normal rain water should be. They agreed with our methods though, and said pH would depend on pollution.

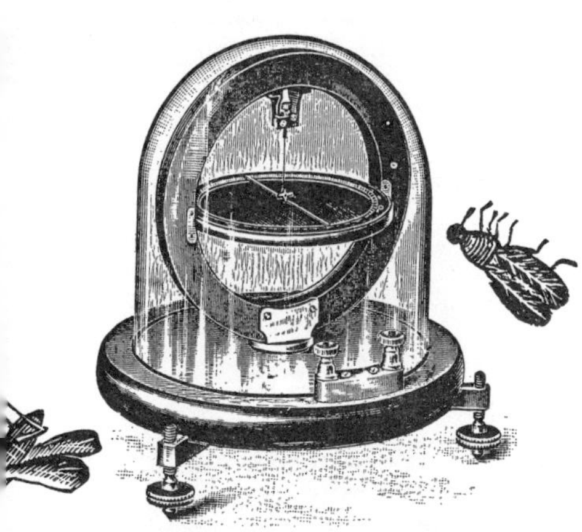

HIERONYMUS MACHINE

We received an article clipped from **Saga** Magazine describing a "Psionic" mechanism which was called the "Hieronymus Machine," after its inventor, T. Galen Hieronymus. The device is supposed to extract an energy from the surroundings and focus it on whatever target the operator desires, with the result that some seemingly impossible results were often obtained. Details of the design and principle can be obtained from its patent (#2482773). The machine consists of a number of knobs attached to electrical rheostats and a compartment in which various samples of minerals or other items could be placed. Dials are turned till the operator "senses" the correct position by rubbing a plastic sheet till it seems tacky to his touch. Under the plastic is a coil of wire connected to nothing at all. Some experimentors think the machine does "something." Others say it only increases the confidence of the operator, who must himself be a psychic.

When a McGraw-Hill executive heard that a modified version of this machine could kill insects at a distance, an experiment was tried with caterpillars in three fruit trees. Some caterpillars, a photo, and a piece of the tree were sent to the machine operator (Dr. Hieronymus) in Florida from 1,200 miles away, and only the center tree was "treated" for three days. The other trees were not treated.

Within the three days all the insects were gone or dead on the "treated" tree, without the other two being affected. This was accomplished by putting strong poison on the photo, which was then inserted into the machine. Another experiment was done in Pennsylvania using poisons on a photo of Japanese beetles on farm crops, whereby only some of the rows were treated on the photo. 95% of the insects died in the rows that were "poisoned," whereas other rows shown in the photo but not treated, were not affected.

THE HIERONYMUS MACHINE seems to be reality (or so I am told by a reliable person who claims to have seen it). He has given me additional information (including the patent), and it seems that what is described in the patent is possible and amazing, though the patent does not claim that it can do what the article describes; it does, however, prove the existence of such a form of energy. TO FIND OUT THE TRUTH, I am trying to telephone the inventor and arrange to go and see the machine and/or get one.

On the afternoon of January 8, 1971, we took off in a friend's VW and drove straight through to St. Petersburg, Florida, arriving Saturday night at the home of Dr. Henry Mandel, where we visited and spent the night. For some time, Dr. Mandel has been working in the field of mental, emotional, and physical disorders which are due to spirit obsession. Unfortunately, on the weekend we were there, the medium with whom he works was not present, so we were not able to observe the entire procedure, but we heard tape recordings of the seances in which obsessing spirits were induced to speak through the medium and describe their situation. We could also hear the way in which these spirits were then encouraged to leave the obsessed individual, and proceed to a further understanding of their role in the spirit world. We were also given copies of some of the tapes.

I suggested to Dr. Mandel that he arrange to video-tape the work that is going on, and have promised to help him with some technical information. If that works out well, I expect that we will be able to obtain television time and present the work to the public.

On Monday evening, we arrived at the home of Dr. Hieronymus. Having already promised to spend that night at the home of my parents in Miami, we could only spend an hour or so with him, and we arranged to return for a longer visit on the following evening. But when we arrived that night, Dr. Hieronymus apologized for the fact that he had made a mistake and forgotten that that day was his wife's 70th birthday, and that he had previously arranged to take her out to a restaurant for dinner and that the table was reserved, and he could not let the occasion pass. (Dr. Hieronymus is 75, but he acts and looks like a man of 50.)

As a result, we could only spend an hour with him that night. We asked if we could delay our return home and come to see him again the next morning, but he did not seem to feel it worthwhile. It seemed that Galen is not eager to give out all the information about the instrument as possible, though we did see it work and tried to operate it. It is not possible, apparently, for everyone to operate this instrument, but a certain psychic sensitivity is required. Some people can develop it quickly, and some not so easily. Some may never develop the ability.

We did discuss the **Saga** article, and were told that it contained a number of serious inaccuracies, and a description was given as to how the experiment was actually done. No poison was used, but only oil of cedar. First, some live worms from the tree were tested with the instrument, and then various chemicals and essences were inserted in a vial along with the worms until it was found that oil of cedar greatly lowered the vitality of the worms, but it would not harm the tree. We were told that by properly using the instrument, it was possible to beam the energy from the oil to the tree, thus lowering the vitality of the worms so that they left the tree or died if they were unable to leave.

We could only speculate as to the reasons why Dr. Hieronymus was not so eager to give us more time and information (though he was not unfriendly or secretive). It may be that he is fearful of harrassment (which many users of these devices have suffered), or he may be fearful of the devices being misused. All I can say is that he seemed to feel that the time for publicity was not yet at hand.

He was familiar with the book, **Psychic Discoveries Behind the Iron Curtain**. He felt as we do that that book will soon lead the West to a breakthrough in the field of psychic sciences. I would like also to advise anyone who has not read it to do so soon.

Altogether, we saw the Hieronymus instrument in use by two operators in two different places, and I am convinced that it is not a hoax. We have enough information (from the patent, mainly) to construct the device, if a great deal of time and effort were put into it. I believe it should be done, but maybe the time is not yet right. However, I have been told that AT&T as well as IT&T are experimenting with such equipment.

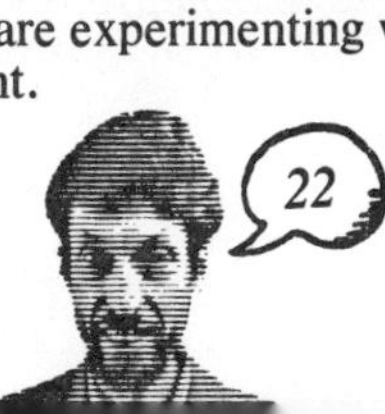

THE HUMAN DIMENSIONS INSTITUTE is a New Age center for the study of spiritual growth combined with sophisticated scientific research.

"We must reorient our concepts of reality to include an appreciation and experience of 'wholeness'—the interrelatedness of physical, emotional, mental, and spiritual energy systems. This requires us to start thinking and acting in terms of 'wholes'—our own individual energy systems mutually interacting with the energy systems of the total environment."

The above paragraph is an expression of their guiding philosophy. A full article on their work appears in **Psychic** magazine for March/April of 1972. The HDI is located at Rosary Hill College, Buffalo, New York. Eighteen thousand people have attended HDI lectures in recent times.

ARF TO GET A CHART RECORDER.
This is an instrument that records electrical variations in graphic form on a chart. It produces a permanent record of experiments in which brain waves, or the resistance of plant leaves, or anything else, are affected by material or non-material causes. Such instruments are generally very expensive, but a friend has gotten one for very little from surplus. It is a modern portable unit that can make two graphs simultaneously so that there is a visual coordination between two things possible. For instance, one for plants or animals, and the other channel might record what changes the stimulus produces in the electrical resistance of another plant.

The recorder can also be used to chart brain waves or skin resistance and many other phenomena, and should open up some new possibilities for original scientific research.

JONATHAN LIVINGSTON SEAGULL is a book by Richard Bach (Macmillan) that everyone should read. It is a short story about birds and flight—flight beyond all realms of fancy. It is about breaking through all limitations of time and space and consciousness. It is not just for the birds. Contains beautiful photos of seagulls in flight.

NEW PEOPLE—Charlie Moore is here to visit with us for a little while. Charlie has been thinking about the work for a long time and sees things about as follows:

"ARF is working to provide much needed information about the latest scientific accomplishments in the

BY THORN BACON

SUPPOSE you were to be told that the philodendron plant resting on the window sill above your kitchen sink screams silently when you break a breakfast egg in the frying pan, or that the potted drecena on the sun porch grows apprehensive whenever your dog goes by?

Finally, would you dare believe that when you accidently cut your finger the dying cells in the drying blood transmit signals to the philodendron, the drecena and the parsley in your refrigerator?

Provocative questions? Indeed, yes, but ones which are being seriously, soberly and quietly investigated by scientists at several major American universities as a result of some bizarre findings by the Backster Research Foundation of New York City.

The object: To discover if there is an unknown communication link between the cells of plants and animals through which distress signals are transmitted that broadcast threats against any member of the living community!

These staggering implications were reported in an abstract published on September 7, 1967, by Cleve Backster, a former interrogation specialist with the Central Intelligence Agency, who operates a New York school for training law enforcement officers in the techniques of using the polygraph – commonly known as the lie detector.

Backster was one of a four-man panel of experts called to testify before the 1964 Congressional Hearings on the Use of Polygraphs by the Federal Government. Following duty with the CIA as an interrogation specialist, he became director of the Leonarde Keeler Polygraph Institute of Chicago. Since 1949, he has acted as a consultant to almost every government agency which makes use of the polygraph. He introduced the Backster Zone Comparison polygraph procedure, which is the technique standard at the U. S. Army Polygraph School.

Changed his life. Teaching polygraph, however, became a secondary interest to Backster on a February morning in 1966 when he made the discovery which changed his life.

These are the words he used to describe what happened in his laboratory that morning:

"Immediately following the watering of an office plant, I wondered if it would be possible to measure the rate at which water rose in a plant from the root area into the leaf. I chose the psychogalvanic reflex (PGR) index as a possible means of measuring the rate of moisture

"Staggering as it may be to contemplate, a life signal may connect all creation..."

ascent. The pair of PGR electrodes could be attached to a leaf of the plant. Hopefully, by using the Wheatstone bridge circuitry involved, I could measure the increase in the plant leaf's moisture content onto the polygraph tape.

Continued →

areas of altered states of consciousness, healing, spiritual phenomena, etc. I plan to spend a good deal of my time here studying these subjects, and hope to be able to share some of my personal discoveries with all of you.

"As I see it, however, ARF has a purpose far beyond that of serving as an information center. Our main task would seem to be to use all the resources and problem-solving techniques at our disposal to find peaceful ways of bringing about the tremendous changes in human relationships which must come about if we are to survive. We have a difficult job to do and will need a lot of help.

"Recently, I began studying with a highly unusual educational institution called the University of Science and Philosophy. It was founded by Walter and Lao Russell to teach creativity through knowledge of the **cause** of original thinking. The founders write:

" 'The pattern of human education has been founded upon the building of mechanical humans who must be fitted into our material civilization as operating parts of a material system. There has never been an adequate institution devoted solely to the nurture of the genius in man.'

"The school has no residents but teaches through a one-year, home study course. The course has helped me greatly to understand the cause of psychic and physical phenomena not adequately explained elsewhere. Later lessons promise to cover Walter Russell's controversial explanations of the cause of interrelationship of matter and energy. The teachings are strictly scientific, but should help the student tremendously to reconcile the seeming conflict between science and religion.

"For a short, fascinating account of the life of Walter Russell, read **The Man Who Tapped the Secrets of the Universe** by Jeffry Furst.

"Inquiries about the University of Science and Philosophy should be addressed to them c/o Lao Russell, Swannanoa, Waynesboro, Virginia.

"I would also be happy to share my experience with their work with any of you who call or write."

VIETNAM. No doubt all of you are very much concerned about the war, and so am I. One of the reasons that I have not written more about it in the past is that the work of ARF is supposed to be non-political.

However, that does not prevent us from being concerned and using the

"Deciding to pursue the idea, I placed a psychogalvanic reflex electrode on each side of the same leaf of the nearby *Draena Massangeana* plant with a rubber band. The plant leaf was successfully balanced into the PGR circuitry, its electrical resistance falling within the resistance limit of the instrumentation.

"Contrary to my expectation, from the outset the plant leaf tracing exhibited a downward trend. Then, after about one minute of chart time, the tracing exhibited a contour similar to a PGR reaction pattern typically demonstrated by a human subject experiencing an emotional stimulation of short duration. Even though its tracing had failed to reflect the effect of the watering, the plant leaf did offer itself as a possibly unique source of data.

"As I watched the PGR tracing continue, I wondered if there could be a similarity between the tracing from the plant and a PGR tracing from a human. I decided to try to apply some equivalent to the threat-to-well-being principle, a well-established method of triggering emotionality in humans. I first tried to arouse the plant by immersing a plant leaf in a cup of hot coffee. But there was no measurable reaction.

"After a nine minute interim, I decided to obtain a match and burn the plant leaf being tested. At the instant of this decision, at thirteen minutes fifty-five seconds of chart time, there was a dramatic change in the PGR tracing pattern in the form of an abrupt and prolonged upward sweep of the recording pen. I had not moved, or touched the plant, so the timing of the PGR pen activity suggested to me that the tracing might have been triggered by the mere thought of the harm I intended to inflict upon the plant. This occurrence, if repeatable, would tend to indicate the possible existence of some undefined perception in the plant."

Backster began to explore how the suffering of other species affected his plants. He bought some brine shrimp, ordinarily used as live food for tropical fish, and killed them by dumping them into boiling water. As he saw the polygraph recording needle leap frantically, he was awed by a startling and apparently new concept: "Could it be that when cell life dies, it broadcasts a signal to other living cells?" If this was so, he would have to completely automate his experiments, removing all human elements which might consciously or unconsciously contaminate the results.

Space age lab. In the three years since, Backster has spent many thousands of dollars in transforming his offices into a space-age assembly of mechanized shrimp-dump dishes, a sophisticated electronic randomizer and programmer circuitry and multiple PGR monitoring devices. But the results continue to point to a capability for perception in all living cells – a perception

Cleve Backster's polygraph says plants feel apprehension, fear, pleasure and relief. Has he found an unknown kind of communication that links all living things?

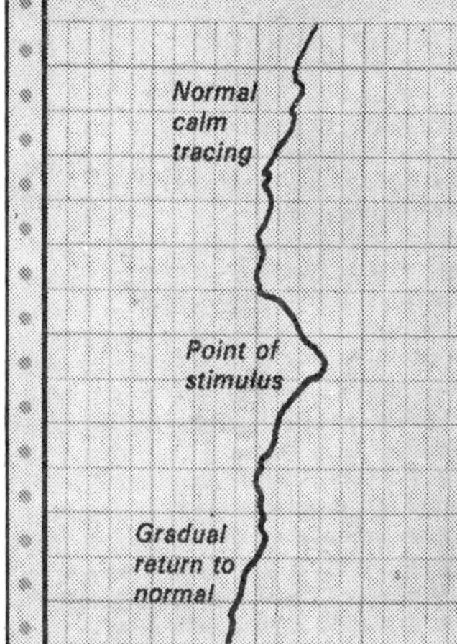

Above, a typical PGR (psychogalvanic reflex) reading from a polygraph of a human test subject experiencing a deliberate emotional stimulation. The chart passes under the stylus at the rate of six inches per minute; each horizontal division represents five seconds.

Below, Backster's first reading from a plant, a *dracena massangeana* being watered on February 2, 1966. The upward tracing in the center of the chart was his first suggestion of some type of arousal reaction from the plant.

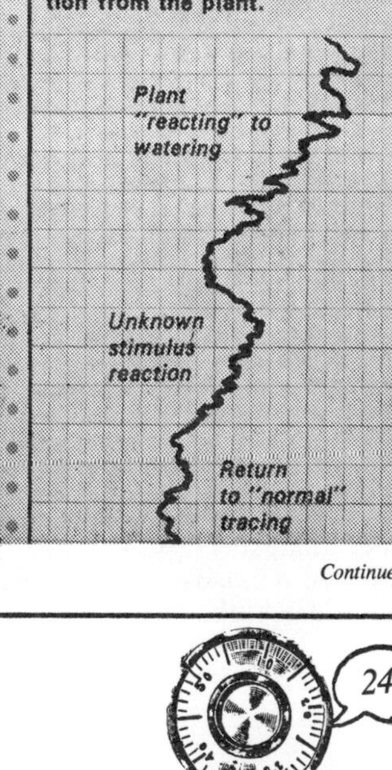

Continued →

mind control method to seek some answer as to how the war could be ended. In fact, I had a strong feeling that I should do this at one point, and so after talking with Charlie M. about it, we decided that we should try to use psychic methods to help toward a solution.

At a deep level of mind, I received a message which seemed to offer an acceptable way out of the war as well as a warning of what would tend to happen if a cease fire did not take place soon. It seemed clear that the message was one that should be given directly to President Nixon rather than published, and I am still not certain if it would be good to publish the contents of the message itself. I tried to get the message to Nixon directly by telephone, but was advised that I could only give it to the White House aid, and he would not be able to let me know whether Mr. Nixon had received it.

The Bell Telephone operator who heard my first attempt to get to the President (person to person) suggested calling the FBI. I tried that, and they suggested that the Secret Service was the proper authority for psychic messages concerning the President. I called Secret Service, and was told that they had the ear of the President, and suggested that I let them record the message, and that if I wished, I could also send it directly by post.

Half an hour after the Secret Service recorded the message, they called me back and told me that it had been given to someone in the White House who would take it directly to the President. I sent confirming letters afterwards. As yet, however, there does not seem to be any cease fire, and, in fact, I have heard talk of allowing Hue to be completely destroyed rather than allowing it to fall into Communist hands. I hope that all of you will do whatever you can to help a cease fire come about.

MOON MONUMENTS—I have now seen a copy of one of NASA's original photos of the "monuments" and it still looks genuine. I have sent for some of the original pictures to follow it up, and will report to you if more and better information is obtained.

AIR WAR IN VIETNAM. Few people are aware of the way that the Vietnam war is being carried on today. The soldiers are being brought home, but the killing of poor people is not stopping. Instead of soldiers, very sophisticated methods have been developed to detect the presence of liv-

that Backster calls "primary". I asked him for more details:

Q. What do you mean by primary?

A. I mean primary in the sense that this perception applies to all cells that we have monitored, without regard to their assigned biological function.

Q. What types of cells have you tested?

A. We have found this same phenomenon in the amoeba, the paramecium, and other single-cell organisms, in fact, in every kind of cell we have tested: fresh fruits and vegetables, mold cultures, yeasts, scrapings from the roof of the mouth of a human, blood samples, even spermatozoa.

Q. Do you mean that all of these cells have a sensing capacity?

A. It seems so. Incidentally, we have tried unsuccessfully to block whatever signal is being received by using a Faraday screen, screen cage, and even lead-lined containers. Still the communication continues. It seems that the signal may not even fall within our electrodynamic spectrum. If not, this would certainly have profound implications.

Q. What kind of a signal is it?

A. I can answer your question better by telling you what we think the signal is *not*. We know it is not within the different known frequencies, AM, FM, or any form of signal which we can shield by ordinary means. Distance seems to impose no limitation. For example, we are conducting research that would tend to indicate that this signal can traverse hundreds of miles.

Q. Are plants attuned to stress?

A. Perhaps. I used to have a Doberman Pincher in my office. He slept in the back room where I had an electric timer hooked to a loud pulsating alarm, which was located directly above his bed. Actuation of the timing mechanism was accompanied by a barely audible click which preceded the alarm by approximately five seconds. The dog would invariably hear the click, and would leave the room before the bell, which he disliked intensely, started to ring. Although in a different room, with the plants, I knew exactly when the dog was leaving his room, even though I could not hear the click, because the plants acknowledged his movements by showing reaction coincidental to the click, reflecting the Doberman's anxiety.

Q. In the final analysis, aren't you saying that we must re-assess our definitions of sensory perception and intelligence?

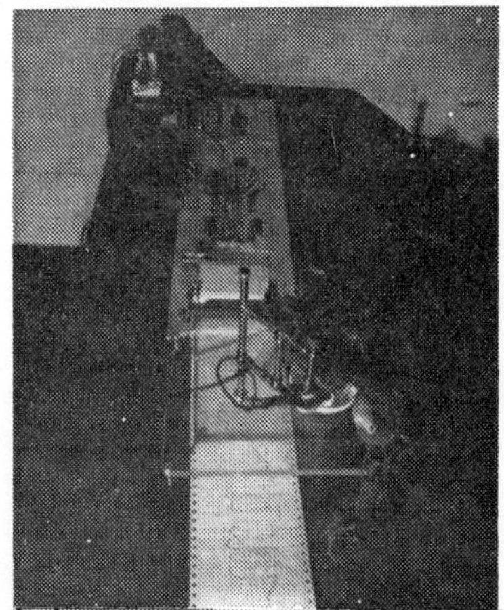

Further tracing from Backster's experiment of February 2, 1966, showing his first deliberate stimulation from a test plant. Backster regards it as significant that his intent to harm the plant produced a stronger reaction (anxiety?) than the actual burning.

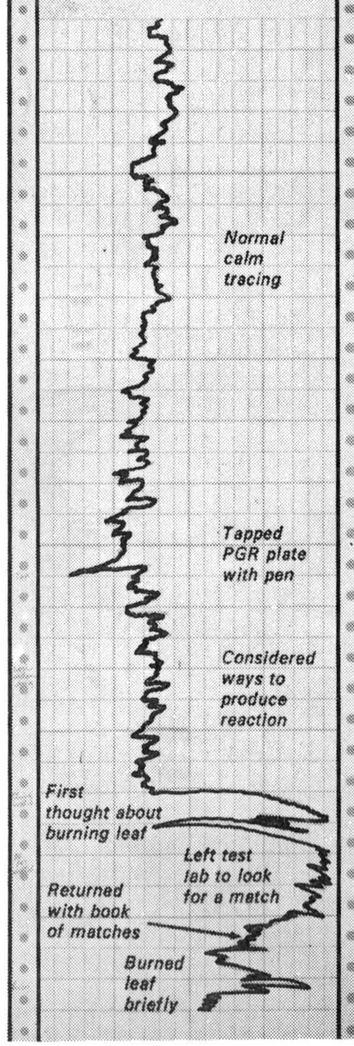

Continued →

ing things wherever they are, and these "people sensors" are dropped all over the countryside. They send signals to high-flying aircraft hovering continuously overhead, and these transmit the signals of many sensors back to the computer centers in Thailand, where the computations are made, and automatically controlled bombers are sent to the location to kill whatever is alive there.

The pilot does not even have to know when or where his bombs will fall; his job being mainly one of bringing the plane back after the bombs have been dropped.

The American Friends Service Committee is funding a group called NARMIC, which can be reached by calling (215) LO3-9372. They can supply you with all the details of this new type of automated warfare, and they have a slide show which is being shown from time to time throughout the country.

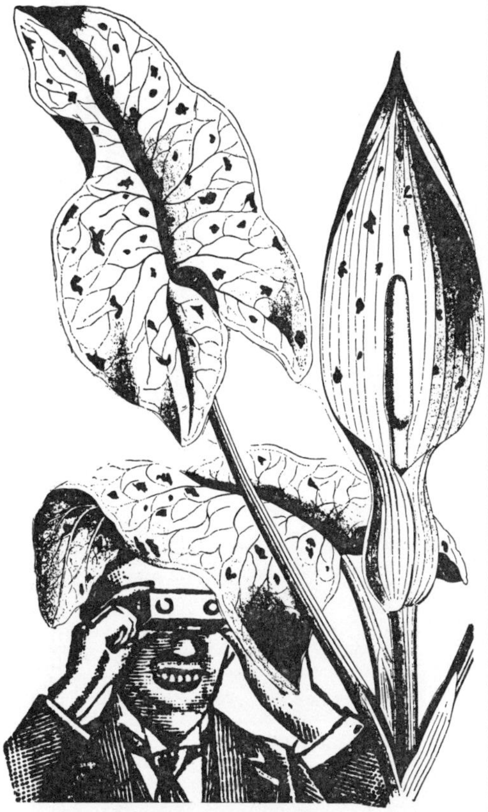

CONFERENCE NOTES

ARF attended a recent conference on Kirlian photography, auras and acupuncture which was extremely interesting, and I hope to be able to follow it up in the near future.

We were able to see numerous brilliantly colored and black and white slides of the auras of leaves and fingerprints, and also to see the equipment

A. Who can say at this point? There are certainly implications here that could have profound effects on those concepts. Our observations show that the signal leaps across distances, as I said before. I have been as far away as New Jersey – about fifteen miles from Manhattan – and have merely thought about returning to my office, only to learn when I returned that at the precise moment I had had the thought – checked against a stop watch – there was a coincidental reaction by the plants to the thought of coming back. Relief? Welcome? We aren't sure, but evidence indicates something like relief. It isn't fear.

Do plants have emotions? The trend of Backster's research results does indeed embrace profound implications. Do plants have emotions? Do they make strange signals of awareness beyond our own abilities to comprehend? It seems so. Personally, I cannot imagine a world so dull, so satiated, that it should reject out of hand arresting new ideas which may be as old as the first amino acid in the chain of life on our earth. Inexplicable has never meant miraculous. Nor does it necessarily mean spiritual. In this case, it may simply prove to mean another extension of our natural laws.

Let me leave you to ponder a question Backster asked me. Many hunters have observed that game animals somehow sense the exact moment of the opening of the hunting season. We can perhaps ascribe this to the noise of the first gunshots. But, how can we explain the similar observation of game's apparent awareness of the exact moment of the season's close? Cleve Backster may be approaching the answer to that question, and a lot of others.

The editors wanted proof... here's what the philodendron gave them!

Editor's note: The editors of National Wildlife *were as doubtful as you may be after reading Thorn Bacon's account of "Backster's Phenomenon", so we visited and photographed him in his offices, just off Times Square in New York City.*

We found a quiet, polite, serious and successful student of the psychology of interrogation, working almost full time on the exploration of his discovery in an office cluttered with extremely sophisticated electronic gear and decor-

Wheatstone bridge circuitry is held in place on philodendron leaf by a special clamp device.

Typical plant reaction to a carefully randomized brine shrimp death, made on the automatic polygraph with no human in the laboratory.

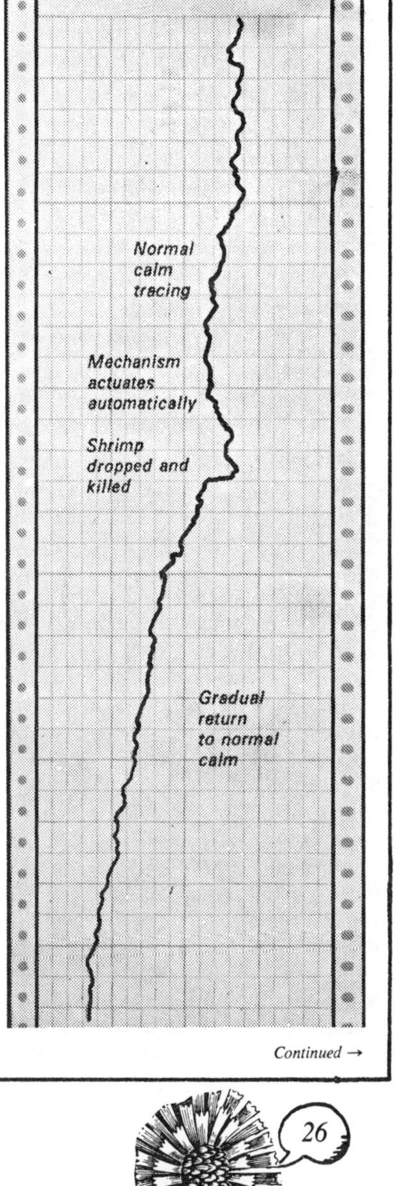

Continued →

with which such photos were made. The equipment was far simpler than anything that I had expected. One of the units consisted mainly of an electronic type automotive ignition system which provided the high voltage pulses necessary to produce the Kirlian effect. Hopefully, in the near future we might be able to supply people with the plans for such a device.

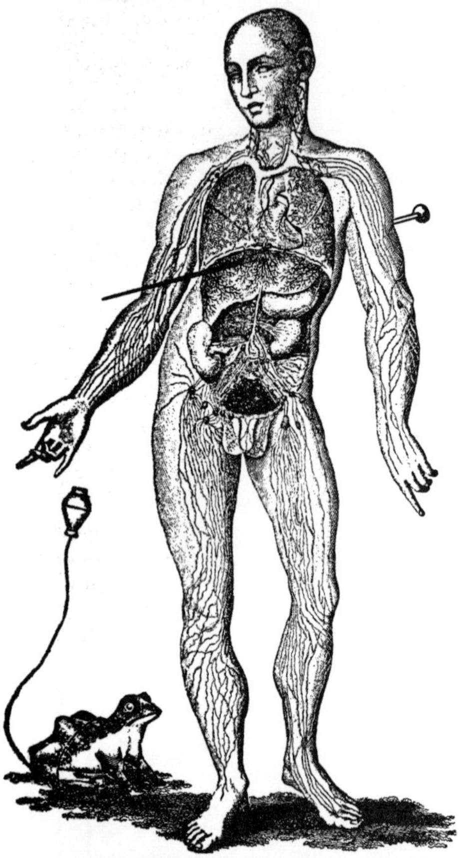

ACUPUNCTURE, the ancient art of Chinese medicine, was also discussed at some length in the conference, and a Russian machine was demonstrated which uses electronics to locate acupuncture points with exactness. When the points are located, it is possible to stimulate them in various ways. This can be done by inserting needles into the points (which is the commonly employed Chinese method), or by massage, or by French device called a "sonopuncture," which stimulates the points with ultra high frequency sound. That last seemed most interesting, and we are hoping to return to New York City and see that instrument, possibly in a demonstration of some sort. The tobascope (Russian instrument to detect acupuncture points) also seems to be a relatively simple unit to construct, and we hope to be able to obtain the plans for that.

ated with thumbtacked records of plant – and other cell life – reactions.

He showed us the original tape from his first discovery of the phenomenon, and yard after yard of tapes from succeeding experiments. One thing impressed us immediately: First, Cleve Backster is not some kind of a nut. He really knows his business, and is pursuing his investigations with great care to avoid any chance of criticism from the doubting scientific community, though he admits that seems inevitable.

As we talked, Backster set up his specially modified polygraph with a fairly ordinary philodendron leaf clamped in position for reading the psychogalvanic reflex index. He mentioned that he no longer handles his plants with anything but great care, since they seem to be attached to him as their owner and caretaker. When a plant must be handled or stimulated to produce a reaction, that is done by his assistant, Bob Henson, who "plays the heavy".

As we sat, chatting, the pen traced a graph of normal repose for the plant, until Bob walked in the room. The graph turned suddenly to one of agitation, and bobbed markedly until he left. Then it calmed down again to a normal tracing.

Later, we talked about ways to stimulate the plant for a photograph, and Backster explained that he preferred not to "hurt" the plant. I remarked that perhaps I could do it, and reached for a match, watching in astonishment as the plant produced a violently agitated reaction even as I began to speak.

Still later, the plant's readings became calmer and calmer, and Backster explained that after an extended time, they seemed to become accustomed to stimuli and their reactions became less marked. At that point I blew a cloud of cigarette smoke over the plant without warning, and it produced a jagged little graph that Backster didn't try to interpret but which I proclaimed to be a reaction of annoyance.

While George Harrison was shooting the photograph that appears on pages 4 and 5, Backster suddenly asked him if anything was wrong; the plant was showing something like a sympathetic reaction to consternation, but was not being stimulated in any way. George admitted that he had just discovered that one lens was not working properly, and had been worrying about the photographs he had already made.

Altogether, we ran the machine on that plant for two hours, and produced a dozen very interesting reactions, some of which Backster recognized (though he is very reluctant to try to interpret them in human terms) and some others that made no particular sense at all, like the up-and-down reading yielded from a telephone conversation Backster held in a neighboring office. The plant reacted differently to the periods of Backster's talking and listening for some reason. But it did react.

So the reactions continue, and Cleve Backster's work continues, as he attempts to analyze the nature of the plants' graphs. Some of the possible applications of the phenomenon, in medical diagnosis, criminal investigation and other fields, are so fantastic that he asked me not to repeat them here. His first serious paper on the phenomenon, titled Evidence of a Primary Perception in Plant Life, *is scheduled for publication in the* International Journal of Parapsychology *in January 1969. He awaits the reaction of the scientific community; we await the reaction of* NATIONAL WILDLIFE *readers. What do you suppose he has discovered?*

DICK KIRKPATRICK

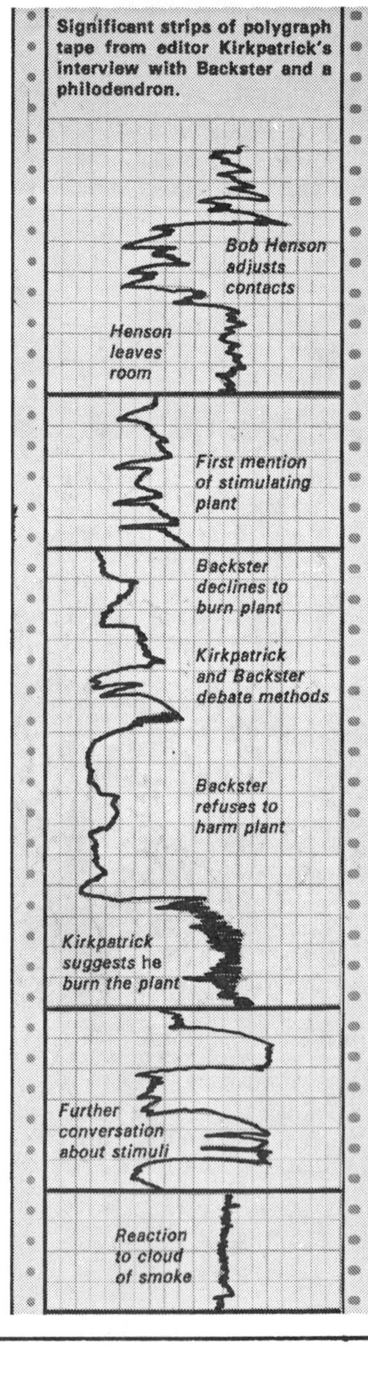

Significant strips of polygraph tape from editor Kirkpatrick's interview with Backster and a philodendron.

NEW IDEAS: College of Communes: How about starting a kind of evening school for people interested in starting or joining communes. "Teachers" will be recruited from existing groups, and visits to communes arranged. Classes will take place at commune sites. Lets hear from communes (and others on this).

"Rent a Hippie" might become the name of a new commercial company to bring "stright" and "hip" folks together. Ads in suburban papers offer to bring hip characters to homes of other Americans (who could invite friends), and the visitor would be prepared to answer all questions about the new life styles, drugs, communes, etc. Fee would be arranged to cover ads and living costs of hip characters so engaged.

LADY is in LOVE; this time with a German shepherd who jumps a 42" fence!! Sends warm greetings and accepts wedding presents. Also reservations for puppies???

ARF AND COMMUNES

The Aquarian Research Foundation was started in the hope of finding ways to overcome society's resistance to social change. After twenty years of communal living and full time work in peace-oriented and radical groups, I have come to feel that there is a tremendous pressure for social change, and at the same time a greater resistance to change (which is due to fear and ignorance of people who do not understand what it's all about).

In order to help the New Age to come quickly and without great suffering, I have tried to get together a group of people who would live communally in order to research ways of blowing people's minds (without drugs), and so help to open them up to the fantastic discoveries that are now being made, and which can lead to the tremendous changes which are needed.

In the 18 months since we started here, many have helped me, and I have been able to uncover and publish information about research in new forms of energy (plants' sensitivity to thought), the Hieronymus experiments, astrological methods of determining human fertility and sex of unborn children (astrological birth control), and many evidences of the permanent nature of the life of each individual.

But in spite of the help of many different people near and far, the ARF has not yet become a commune or group of people working together on the same project. Jeff and Ronni are here with me, having come from California in the hope that such a commune would now form. At present, Jeff is writing a part of this newsletter dealing with our personal struggle to form such a commune. It's not easy, and we have had to see that we all have a lot of faults.

Ronnie is cooking and baking bread and also reading the translation of **The Life of Johann Christoph Blumhardt**, which tells how the strange illness of a girl named Gottliebin Dittus in 19th century Germany led to the discovery of a very real case of spirit possession, and thence a completely new look at the New Testament to find out what Jesus really meant. As a result, Blumhardt came to have a tremendous gift of healing for thousands, and a deep understanding of the meaning of the Kingdom of God—the rulership of love—which the New Age is all about. Hopefully, in the next few days, Jeff will finish the reprinting of that book.

THE AMERICAN BOOK OF THE DEAD is a book which we are still working on and hope to complete in the next few weeks. With regard to that book, we have had a very excellent offer from a small publisher in Amherst, Wisconsin who has offered to print us 1000 copies of the book free of charge, and keep 1000 copies for his own sale.

SILVA MIND CONTROL is another group that is now teaching people to develop their psychic abilities by helping them to get control of their brain waves to the point of being able to go at will into an alpha brain wave state. I was able to hear a tape of the introductory lecture of the course and had someone here who was taking it. I have also met two others who took it a few months ago. All three affirmed that they were able to develop psychic abilities through the course, but it was also said that if these abilities were not used frequently they are soon lost.

There are groups of people who have taken the Silva Mind Control Course and who meet together regularly for practice sessions. The course is open to the public, takes a total of 48 hours and costs $175 (half price for students). I have been half-promised a scholarship to the course on the basis of my work, but I don't know if it will come through yet.

PSYCHIC INSTRUMENTS may soon be available at a low price. We are working together with an electronics firm to develop an instrument that is capable of measuring the tiny changes in electrical resistance which Cleve Backster found to prove that plants can detect human thoughts.

I have already been testing one sample model of this instrument, and it seems to have at least as much sensitivity as Backster's polygraph. Backster is now publically admitting that he has tested non-living materials (such as conducting rubber and some minerals) and obtained evidence that these can vary their resistance in response to human thought.

Cleve also reported that it was possible to get a pulse from a fertile egg at a rate of 128 per minute, even though the egg had not yet been incubated and no heart (which would later beat at that rate) had yet been formed.

The present instrument which I have been testing has given definite visible reactions which might be due to telepathic messages. In order to ascertain that, it would also be necessary to have a suitable chart recorder to record the variations on a paper strip that moves along at about four inches a minute. Such equipment is ordinarily quite expensive, but I am attempting to find people to produce it at a very low price so that all the equipment necessary for repeating Cleve Backster's experiments can be had for about $100 or less.

It also seems possible that the same kind of energy which is the cause of resistance changes, and which is related to thought, is the basis of the Hieronymus machine; if this is so, it might be possible to get the machine to produce visible indications on a meter instead of being only usable with a finger-reading psychic. We might be able to make a test of the instrument in that way in the very near future.

RENT A HIPPIE is getting a bit of interest from some people here and there, including someone who is running a radio station in Canada and who called to ask if we could send them recorded program material. I am afraid that I was not able to give them anything because of the lack of time, and that for the same reason I cannot take on any project like the "Rent a Hippie" idea, but I hope that others will take it on.

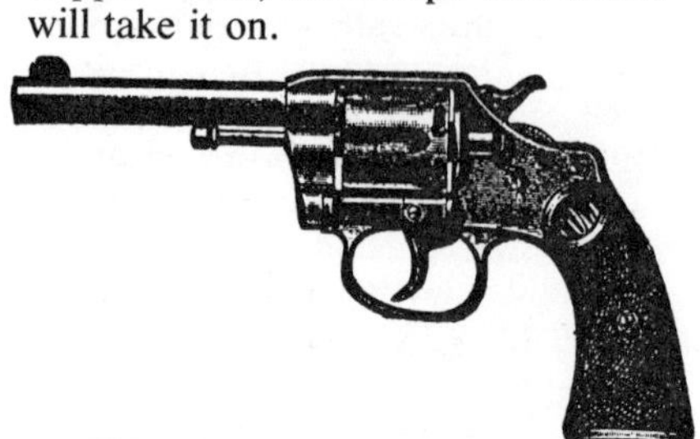

FREE SCHOOL is still something I would like to see started here. We are looking for teachers who would like to work with us in order that the parents would be satisfied that their young people are getting an "education", but we wouldn't actually want to have formal classes. Enough to learn here without that. One 16 year-old is already wanting to come and his father was here to talk with me about it today. No teacher and concern about the boy's "education" were some of his excuses for waiting, but in the end he said he would not hinder the boy from coming (but his mother is another matter).

LADY'S NEXT LITTER OF PUPPIES are due very soon; we are expecting a bunch of little German shepherds any day now. We discovered that we were having difficulty selling them last time because our price of $50 was so low that people were suspicious.

SENSITIVITY TRAINING EXPERIENCE

I asked Jeff to write about the experience we had at the Council for Social Development, which is a group working with sensitivity training as a means to social change. What he has to say now follows:

HERE AND NOW. How do I feel? How do you feel? Can you explain it? Will you please tell us?

We're here now, trying in some ways to relate to each other. "I don't know how we're going to do it" seems to be the general impression. Someone in the group takes "control" and suggests that we all put our feet into the center of the room, "touching." This is all we do. The next suggestion is that which stemmed from a general agreement of relating in the HERE AND NOW. Right on! So we related individually in this manner; good start. We all want to know about each other, and how we feel.

Sensitivity, or, encounter groups are familiar experiences. We encounter (meet) people in a sense much of the time. In this situation, we encounter each other for the implied sole purpose of sensing and dealing or encountering ourselves and each other. Working from "scratch" to see what will "grow." There's a kind of hush; you know what I mean? We are here to create a new and better image of the "human being" in the sense of being more at one with each other.

Methinks—rapping with people, "getting it on"—and being out front is essential. All right, here we go. Art and I "unravelling" in the center of the group, the "center of attention." I feel a verbal and non-verbal defensive mood in the room. Ego, ego, ego it screams. For one reason or another, the people in the group generally agreed that they wanted their problems attended to. They agreed that too much time had been spent on Art's problems, like half the time at many meetings. I feel that there is something at least that everyone in the group could relate to Art.

The generally accepted fact that we have relationship struggles is quite a broad and essential topic of discussion. This is what the two of us are encountering. Now really isn't it only our ego that is depriving us of full benefits? If you would be deeper and pay close attention to what is happening, you might find that **you** are in the center of the circle and just the illusive names and individual situations are changed.

Remember, concentrate on the HERE AND NOW and not just our own egos. For myself, I needed to kick out the jams, throw open the doors. To see the light I needed to do this. It was working too hard on my head. Some of those in the circle let me know where my ego was at. When it was all out of my head and I thanked them, I only hoped that others would have let more of theirs show. There is really nothing to be afraid of but what your ego fears, "that of being discovered." "There is nothing to fear but fear itself." Don't fight, wrestle, and fear your ego by keeping it tight.

When you're familiar with it—your ego—you have a good chance of being friendly to the point of laughing with others who are at the same place, or, if you're alone, laugh like a crazy person. Basically, these people at CSD have a good thing going. Most of the theory as dealt with so far seems very useful. Either that, or people that use it have helped me. I guess, however, that many encounter groups sell short of being sensitive to a group brotherhood.

Encounter is a good practice. It should be continuously on your mind for greatest effectiveness. I feel that ego is obviously the one basic problem to deal with openly. It's basic to everyone and everything stems from ego; if used properly, it will lead to real freedom. All problems stem from misuse of ego, or fear (which is the same thing). Ego blocks development of our real selves.

I recommend the encounter group to everyone. Practically speaking, to those of you who are least willing to be out front. The same things that are hindering you from being honest are hindering us; that is; fear of not being accepted. It helps me by remembering we are all individual expressions of the same things. For more information call CSD (215) WA 5-1256.

After Jeff wrote the above, I felt that I still had a few things to add to it. For some time, I had been taking part in these groups at the invitation of the CSD people, and I had all along had the feeling that I needed help. There must be some reason why people have come to work here but have not stayed very long, and some of the fault must be mine. Jeff's coming was the first time that anyone had been here who really wanted to go through a struggle for unity, and Jeff felt that I was not being "out front" (honest) with them. I thought that I was being completely open about everything. So we brought all this up in the encounter session. It developed that the rest of the group also felt that there was something lacking in my honesty, and they were able to express what they felt the trouble to be.

It seemed evident that I had become so fully involved with ARF as a "project" that I was becoming like part of a machine myself, and was not able to share my own personal needs and fears and feelings with others as fully as I thought I was doing. I was "objective," but not personal. It was like a salesman who is so concerned with the things he is selling that he has no regard for the people to whom he is trying to sell.

Aside from ego trips, I think the difference between us is that Jeff is looking for a really together commune. I guess my pamphlet was not too explicit in pointing out that I was seeing the research project as the first purpose of the group, and that those who wanted to work at it would then live together communally here.

As we try to work together, our personal problems will come out, and we will have to deal with them and seek unity in order to go ahead with the work.

I do not feel that we need to perfect ourselves first, but rather that when the New Age is fully come and the world is more as it should be, then we will be more as we should be also.

PUPPIES may come tonight, March 2, 1971. Lady looks really ready tonight.

MIND CONTROL BREAKTHROUGH

The greatest thing that has happened to me in a long time was the experience of taking the Silva Mind Control course. I am now certain that each one of us can develop psychic and clairvoyant abilities in a relatively short time; that all people are naturally clairvoyant, but the kind of culture we live in, and the education we receive inhibits those abilities instead of developing them.

After years of research into the area of the mind, José Silva has developed a course lasting only 48 hours in which we are taught to control our own minds so fully that problems such as nervousness, insomnia, headache, and the like can be controlled and eliminated in ways that give complete relaxation and improve health and mental abilities generally.

My own experience in the class was different than many, but not unique. Knowing that clairvoyance is real, I still did not believe that I could experience it myself. As we finally came to the time of testing at the end of the course, I felt sure it would not work. I tried very hard to get the visualizations that were expected, and couldn't get it. Still, I was, nevertheless, in a hazy sort of a way, getting some accurate information. Then an instructor came over to help me, and still I was not able to see clearly what was happening to the distant person whose illness I was attempting to psychically perceive. Finally, I realized two things.

Visualization can be hindered by trying too hard, and by not caring enough about the person you were trying to help. Later on, I was able to get a few very clear psychic visions of things that people were suffering from at a distance, and they were remarkably accurate.

It is evident that if we are given a few more years of peace in which to work at these things, mind control courses will become part of all school experience for children, education will be revolutionized; the New Age speeds on!

The Silva course was an intensely positive (but relaxing) experience, but now I have to work with the results.

From our minds, we send out smoke and fire. From the fire we get warmth and light, but the smoke is no help at all. We must experience that we are far too important to waste our time playing with the negative aspects of our own and others' personalities, and instead, we must work with all that is positive. I am finding this concept a tremendous help in relating to other people.

And yet, I find myself in the difficult situation of learning to be warm and loving to each individual with whom I come into contact, and, at the same time, to give my full energy to the much larger question of finding out how we can end this age of war and bring in the time of peace and joy. There must be an answer which combines these two concerns in a really loving and positive way, and I have to renew my own personality in order to find and incorporate that answer.

BACKSTER ON TAPE: A full tape recording of one of Cleve's recent lectures is also available to those wishing to hear it. Telephone to arrange a time or to request a copy of the tape.

LADY has given birth to German shepherd puppies again, and in about a month we hope to have eight of them for sale. She is a very gentle dog, good with children, and yet a fine, large watchdog too. They should be ready between April 15th and 30th.

DREAM THEORY IN MALAYA
By KILTON STEWART

Kilton Stewart was a research psychologist who spent a year among an isolated jungle tribe in Central Malaya. The Senoi can truly be thought of as "tappers of man's reflective thinking," for they demonstrated a creative process known to us all but used by few: much of their life style is determined by their dreams. Stewart found that Senoi needs for food, shelter, and clothing were met in a manner similar to most primitive societies. Their language and physical characteristics reflected relationships to peoples of Indonesia, Indochina, and Burma. But it is reported that the last two or three hundred years have been free of internal or external conflict.

Freedom from outside intervention may be attributed to scare tactics, as the Senoi have developed a reputation for using black magic (which does not seem to be the case).

To explain how they have avoided violent crime, armed conflict, mental and physical disease, and eliminated the need for jails and police forces, however, we have to look at the social psychology advocated and practiced by the entire tribe.

Fifteen years of experimenting, says Stewart, with Senoi ideas has convinced him that everyone, irrespective of their actual cultural environment, could profit by learning about them.

Senoi psychology falls into two categories. The first deals with dream interpretation; the second with dream expression in the agreement trance or cooperative reverie. The cooperative reverie is not participated in until adolescence, and serves to initiate the child into the states of adulthood. After adolescence, if he spends a great deal of time in the trance state, a Senoi is considered a specialist in healing, or, the use of extrasensory powers.

The children are seriously encouraged to discuss and analyze their dreams with parental and community help. Dream experiences are related to the child's own psychological makeup. The general theory is that a man develops hostile images of his environment, turning against himself and his fellow men. This hostility manifests itself in physical, psychic, and organic tensions. Dreaming enables him to analyze his psyche and avoid abnormal attitudes toward his fellows.

However, this sytem of interpretation has broader implications beyond personal development. It is felt that positive images in dreams are contributions the dreamer must share with his society. Any pleasures in dreams must be enjoyed to their fullest, and understood as an indication of the favor of beings from the spiritual or emotional universe. Sexual love (in dreams) must be culminated after orgasm, with the dream lover giving the dreamer a poem, a song, or a dance or any useful knowledge that will enable him to express the beauty of the relationship to his tribesmen.

Injury to a friend in a dream requires the dreamer to compensate for such hostility in his waking hours by expressions of friendship and cooperation. Hostile dream characters only use the images of those persons towards whom the dreamer himself shows little good will. If the dream character is a friend who does some misdeed to the dreamer, he must tell the friend so that whatever conflict exists in the relationship might be corrected.

Accepting a friend's dream as being part of his free inner expression is an acceptance of that friend.

After considerable acknowledgement and analysis (and full development) of the dream process, the Senoi learns to neutralize the unpleasant dream experiences and finally reverses them entirely, turning them into constructive contributions for the good of the whole society. He overcomes the bad by adding a good image; then he eliminates the bad entirely by creating a positive product, be it a lyric poem or a solution to a community problem.

The elements of the interpretive process may be seen as: 1) the child receives social recognition and esteem for discovering and relating what might be called an anxiety-motivated psychic reaction. This, among the Senoi, is the first step towards convincing the child that he is acceptable to authority even when he reveals how he is inside.

2) The working of the child's mind is described as rational even when he is asleep. To a Senoi, it is just as reasonable for a child to adjust his inner tension states for himself as it is for a Westerner to expect him to do his homework for school.

3) The interpretation characterizes the force which the child feels in the dream as a power which he can control through a process of relaxation and mental set; a force which is his as soon as he can reclaim it and learn to direct it.

4) The Senoi education indicates that anxiety is not only important in itself, but that it blocks the free play of imaginative thinking and creative activity to which dreams would otherwise give rise.

5) The method establishes the principle that the child should make decisions and arrive at resolutions in his night-time thinking as well as during the day, and should assume a responsible attitude towards all his psychic reactions and forces.

6) It acquaints the child with the fact that he can better control his psychic reactions by expressing them and taking thought upon them, rather than by concealing and repressing them.

7) It initiates the Senoi child into a way of thinking which will be strengthened and developed throughout the rest of his life, and which assumes that a human being who retains good will for his fellows and communicates his psychic reactions to them for approval and criticism is the supreme ruler of all the individual forces of the spirit-subjective-world-whatsoever.

Finally, Stewart concludes by saying that we Westerners feel that the thinking we do while asleep is childish, muddled, or psychotic because we can't accept dreams as significant or creative. And this means a poor education.

The above article is a review of a longer article found in a book of readings called **Altered States of Consciousness**. That book is edited by Charles T. Tart of the University of California at Davis. It is published by John Wiley and Sons, Inc. (New York, London, Sidney, Toronto).

CLAIRVOYANCE AND ALPHA WAVES

About research into Alpha brain waves and biofeedback: Dr. Elmer Green is director of research at the Menninger Clinic and has been doing some amazing work in the field of biofeedback. Dr. Green (and his wife) both realize that we have a rising pressure for change as well as a rising resistance to it, and he saw his own work as an attempt to do something about that. He was fully aware of the danger of our present situation. I also found that he was well aware of many of the other discoveries that the spiritualist movements have been coming up with over the years.

He was also aware of the work of the Silva Mind Control people, and that they have been able to produce clairvoyant states, but was of the opinion that those states occur in the Theta rather than the Alpha wave area.

On speaking with the Vice-President of Silva Mind Control re Dr. Green's comments, Jim Needham told me that the exact brain wave frequency at which clairvoyance is most likely to occur is dependent on the development of the individual psychic. Some experienced and trained psychics who have had a lot of practice can even demonstrate clairvoyance when in the common brain wave state of full wakefulness with their eyes open. Dr. Green told me that he has not been working with clairvoyance himself.

PSYCHIC BIRTH CONTROL!

It seems that with the new discoveries being made in the field of voluntary control of internal organs, and the work of Silva Mind Control, various people are beginning to realize that there are also psychic methods of controlling pregnancy which seem to be quite effective.

In fact, I have been recently meeting women who tell me that it is possible to prevent pregnancy simply by being sure that it is not going to occur, and that no birth control of any kind is required beyond that certainty. On hearing these stories from people who have had children when they wanted (and only then) in spite of known fertility of both partners, I am reminded that certain primitive tribes have believed the same thing for many years, and have always used such methods but have been unable to convince the anthropologists of those facts. The University of Pennsylvania Anthropology Department is helping us to locate more exact information in this area.

MIND CONTROL TEACHER IN PRISON! Today we had a call from one of the people who will soon be very active in the work of prison reform in the State of Pennsylvania. He told me of amazing plans to introduce the Silva Mind Control Method into the prison system and of his hopes to have the course given to the prison inmates by people who are former convicts themselves. This looks like a very promising possibility, especially when we realize the similarity between our prisons and our public schools. If we can have it in the one, we can surely have it in the other.

"KUNDAR'S VENTURE"
By ANNE MEYERDING

I once knew a kindly little dwarf by the name of Kundar. Kundar was a good fellow, always glad to help out a friend, but over a period of time he got to feeling restless and decided that he wanted to go out into the world and find out what was happening. The way he described this feeling was that he knew all about how the other dwarves around him lived, he knew the comforts of home and friends, but he wanted to know if there were other ways of living and maybe even different kinds of creatures. Also he wanted to see new countryside, to see whether all the world had the same hills as his homeland, the same sweet-smelling flowers, the same small sturdy trees, and birds that sang through the day.

After some delay Kundar set off, knapsack on his back and walking stick in hand, to see the world. In several days of travelling he had made new friends, talked with a beautiful lavender flower who called herself Iris, and he had found a small yellow ball. He hadn't known at first what the ball was, but one of his new friends had told him it was a ball and showed him that you could throw it up in the air and catch it (if you were quick) when it came down, or you could roll it along the ground (and knock something over if your aim was good); you could bounce it up and down or, if you had a willing friend handy, you could throw the ball back and forth. Kundar loved his ball and often tossed up as he walked along and caught it when it came down. He became so attached to it, in fact, that he usually carried it in his left hand as he hiked along, and squeezed it now and again.

Kundar kept travelling, meeting new creatures, new grasses—some grew high and golden and bowed before the wind—new flowers and trees, and generally enjoying himself immensely. Other dwarves he met were usually friendly and generous, inviting him to share a meal or to sleep in their hammocks, and he never was hungry for long, but the other dwarves questioned Kundar about his travels and seemed not to think kindly of his being so far from home.

"Don't you know you should be growing turnips and beets and yams at home so dwarves don't go hungry?" they asked.

"But I don't know any dwarf who is going hungry."

"Well, you should grow the things anyway and store them because you never know what might happen tomorrow."

Kundar didn't feel the way the other dwarves did, but after he had talked with many who said that his travelling was the wrong way to live, he began to question his feelings. He still enjoyed hiking over wooded hills and down through green valleys, but he began to avoid the company of other dwarves because he knew they would make him feel uncomfortable with their questions. The only trouble was that he began to feel very lonely.

One day, as Kundar walked through a meadow on a hillside, feeling the warmth of the sun and smelling the sweet meadow grasses, he came upon something lying deep in the grass. He would have passed it by if it hadn't been lying against a particularly beautiful daffodil which had caught Kundar's eye because it reminded him of the color of his little ball when it had been new. The thing Kundar found was made of wood and had been carved into letters like this: INTEGRITY.

Kundar could read a bit, so he sounded out the letters and said a loud, "Integrity."

"What?" asked the daffodil.

"Integrity. Do you know what it means?"

"No, but whatever it is I don't need it because I am content as I am."

Kundar thought for a bit and decided to try to find a village so he could ask the Wise Dwarf there what "integrity" might be.

"After all, I am travelling so as I can learn about things I hadn't known about before," he thought to himself as he set off.

It took him until the next day to reach a village, and by that time he was feeling very excited. He felt a little silly too, getting worked up over a few letters carved from a stick.

"What if 'integrity' doesn't mean anything at all?" he thought. "I'll just look foolish."

Still he persisted, asking for directions to the home of the Wise Dwarf, and when Kundar met the Wise Dwarf, whose name he was told was Ari, he was glad he had come because this dwarf looked as kind as he looked wise.

After the two dwarves had greeted one another in the customary manner and settled down over cups of steaming grog, Kundar pulled his stick from his knapsack and handed it to Ari.

"Do those letters mean anything?" he asked hopefully.

"They mean 'integrity' " Ari said. "This is a very fine piece of work, Kundar."

"But what does 'integrity' mean?"

"Why, integrity is the most important thing in your life. Integrity means doing what you know to be right and true and sticking to it no matter what anyone else says about it."

"It does!" Kundar exclaimed in surprise.

"Yes. When you feel something is good, stick by what you believe because if you give in to other dwarves' opinions and go against your deepest feelings, you'll lose your integrity, and I assure you if you do that, you will lose the most precious thing you have."

Well, Kundar kept on travelling and he learned a great deal about the world but he wasn't lonely anymore. He spent many happy days visiting other dwarves in their gardens and many cozy evenings chatting at their firesides. When they questioned him about his travels, he replied, "My integrity demands it."

Anne Meyerding, who wrote the above story and who is also a mind control graduate (I believe she got the story in a dream which she programmed herself to remember), is now helping to edit and assemble the chapters of the **American Book of the Dead**.

A WEEKEND AT ALBRIGHT COLLEGE: Last week at this time I was spending a three-day weekend at Albright College, where many people were invited to come and visit and speak to students on the subject of the future.

Talks with the Arrakis Family (a commune) seemed to hold the greatest interest for the students. The thing that was most appealing about them is the fact that they were able to share the problems of communal living and new family concepts in such a free and open way. Their personal difficulties were openly admitted and talked about, but so was all the joy of the living and working together as an intentional family. The Arrakis people have asked me to pass on to you a project of theirs which is intended to help people in all walks of life to find new and more meaningful living situations. They call it the:

FREE PEOPLE's EXCHANGE
C/O ARRAKIS, RFD # 1
JEFFERSONVILLE,
NEW YORK 12478

Dear Friends,

Like lots of people, we're exploring less competitive, more cooperative ways of being. We'd like to create a Northeast regional clearinghouse so that we can all help each other out. We'd like to:

1. publish a regional intercommunal exchange newsletter with limited circulation.

2. publish a regional newsletter on cooperative alternatives (i.e. free schools) for a broader audience.

3. provide immediate information and/or contacts on request. Our main interests right now are in:

—problems and solutions for rural communal groups (ecology, getting money, organization, agriculture)

—problems and solutions for urban communal groups (e.g., food conspiracies and co-ops, communes and tenement rebuilding/repairing, getting money, de-traffic plans, viable small industries)

—people, communes, and movement groups seeking people/places/vocations

—barter/exchange/mutual aid needs and offerings

—actually accessible tools and how to get them; new ways of using tools (doing more with less—including publications)

—gathering of tribes, festivals.

Does this make any sense to you? Maybe not. But consider:

—In Vermont there's a cooperative bus company which services movement groups; around Liberty, New York, there's an honest real estate broker who might help you find a decent commune location at a reasonable price; in Connecticut, Meg Robertson and Brian Heath are willing to share their experiences of the British primary schools with anyone thinking of starting a free school; New England Food Co-op can help you buy and/or produce inexpensive nutritional food; and in the Catskills, there's a VW mechanic ready to share and exchange service and skills.

Two things we'd like from you:

1. Does all this square well with your needs? If not, what would square better?

2. What resources do you have or can you suggest things we can list or refer others to?

We'll be looking forward to hearing from you soon!

Warmly,

Jerry, Judy and Sue of the Arrakis Family

Priscilla, Len and Rick of the Inner College, University of Connecticut

Joel and Chuck from the hills of New Hampshire and Vermont

BEAR TRIBE COMMUNES: COMING HERE!* Speaking about communes and such alternatives, a group started by American Indians seems to be doing the most important organizing on a nationwise basis for people wanting to start living on the land. They are the Bear Tribe, and write as follows:

For for past six months, since October, 1970, an ever-increasing group of people have been working together in Northern California to make the Indian way of life possible again. On the winter solstice, we organized ourselves according to Indian ceremonials, and became the Bear Tribe. We will soon have perhaps 150 members living at our camps or working at one of our projects.

The Bear Tribe is not excusively Indian by race; far from it. We have many Indians among our members (from tribes all over the continent and with all manner of traditional background and degree of blood). But we have many more who are non-Indian, who have come to the tribe because they believe that the Indians can teach a way of life in harmony with the spirit of this land before it is too late.

The medicine of Sun Bear, the Chipewa who founded the Bear Tribe, is to bring land and people together. This he has been doing with all of our help in California. Beyond this, and because disaster may be imminent for the civilization which is occupying this land, the Bear Tribe's purpose is to reach and spread Indian ideas in any way possible, so that other groups

**We're not really certain if the Bear Tribe is still in existence, but the idea is so interesting that we unanimously decided to leave it in.–Ed.*

looking to the land elsewhere on the continent may look to Indian ways for helpful experience.

Our own efforts at living tribally on the land here in California continue to bring fruit; we have had losses and heartbreaks, but in a large part they seem to have given us strength. Recently, for instance, we all but lost our very first camp, which we were half unconsciously thinking of as our home. Through these inevitable pressures from reluctant county governments, our right to the land became vulnerable, and we had to move all but a nucleus away, or lose our right to it. That point, though, is that the other tribesmen are simply being spread out among different camps, and without undue concern, we lived up to our preachings about mobility.

This is a good example of the kind of difficulty we have had to face. Another camp lost its land to the dread subdividers, but almost simultaneously found another site, with fewer acres but more isolation. They lost their gardens and homesites, but seem to have more than enough energy to restore the loss in time for spring planting.

I've written a number of articles about the Tribe, and this is the first in which I've begun with the problems. Often enough, responses to earlier pieces which we've sent around the country have indicated that people thought us to be mysteriously exempt from the difficulties that other communal groups ran into. It is important that we do not give that impression because we are trying to encourage other groups to do something along these lines all over the country and do it now. Any given group may be obstructed, and may have to re-form and try again elsewhere; as an historical movement (the Indians believe), this generation will succeed in returning to nature. If this does not avert the strife that is building, it will at least make it possible for many more people to survive.

People are beginning to live with the land here. A base which is just being established north of here (Sacramento) is a good example of the form we think life must take where possible. A tight group of twelve people, which will grow slightly later, has just finished putting a large garden on a sixteen acre patch which was given to us. This borders on a huge wilderness area, and the tribesmen will now move back into the mountains, sending people from time to time to care for the garden, and otherwise living on the move, depending especially on a growing knowledge of edible plants.

Another base, in heavily populated Santa Cruz County, is faced with a completely different situation. They have a small, fairly isolated area, where they are building earth lodges and planting a garden. But they also use donated garden spots all over the county, where they are working together with various groups from the community. There the effort is to be highly visible and to make contact with great numbers of people.

Other bases are in operation (or soon will be). We have one piece of land on which we will be able to teach the Indian uses of the ocean and shore. The idea is to learn whatever we can from the land that comes our way; the members living on the land will learn as much of the Indian way of life as they can, so that they may become teachers and help to seed tribal groups elsewhere. While this is going on, we are trying to reach others around the country, to let them know of our existence, and to begin to spread knowledge of traditional life within nature. We hope that groups will form and find land where they are; much of what is needed for survival can be learned from nature, and it's surprising how much can be learned in bits and pieces from the community.

We will try to make contact with as many groups like this as we can find, and as soon as possible. We may be able to help with information about native ways of life; it is also important that groups on the land begin practical alliances for barter, to trade skills, and so forth.

To this end, the Bear Tribe is sending a caravan around the country, and continuing until we reach everyone who invites us. For some reason, which we are at a loss to explain, we have had more response from Philadelphia than from any other place outside California, and we are looking especially to your city as a source of many contacts and a number of meetings. Any who are interested, or who can help set up a meeting (of any sort), are invited to contact the tribe at P.O. Box 1222, Sacramento, California 95806. The time to teach and prosper has come.

Carl Turlock
(for the Bear Tribe)

The above suggestion of setting up meetings for Bear Tribe people coming with their caravan is understood to apply to the whole country, not just Philadelphia. People seeking other alternatives and communes may also apply to the following groups for information:

Alternatives Foundation, P.O. Drawer A
Diamond Heights Station
San Francisco, California 94131

Send $1 for commune list or $10 for membership.

Heathcote Center
Rt. 1, Box 129
Freeland, Maryland 21053
has a list of "intentional communities that are open or public"—$.50 if you have it.

New Schools Exchange Newsletter is for you who want to know about free schools. Write:

New School Exchange
Petigrew, Arkansas 72752

Subscription price is $10/year or $6/6 months. Request directory of free schools if you want it with subscription.

Or write:
Teacher Drop-Out Center
Box 521
Amherst, Massachusetts 01002

Subscription to Newsletter and list of new schools, $3. Telephone 617-253-9787.

People seeking jobs or a social change orientation may write:
Vocations for Social Change
8951 Canning Street
Oakland, California 94609
Donation of $1 suggested.

SPIRITUAL HEALING FELLOWSHIPS that do not ask for money can be found by writing:
Order of St. Luke
19 S. 10th Street
Philadelphia, Pennsylvania 19107

NEW INSIGHTS ON CANCER
Detection, Treatment, Prevention

For hundreds of years, some scientists had maintained that scurvy, which used to kill thousands of sailors, was due to poor diet and a lack of vitamin C. Their evidence was consistently rejected until finally tested by the British. But even long after the problem was gone from the British fleet, other countries were too stubborn to accept it.

They called the British "limeys" because they had water with lime juice in it while other men died of scurvy for over a hundred years for not using it.

Could it be that similar things are happening today with regard to cancer? Let me share with you the facts that have come to my attention.

LAETRILE I was told that Laetrile, which I saw in Mexico, is one of the most effective controls for cancer and is actually a natural food product; more exactly, a vitamin. It is the same as vitamin B-17, which is also known as amygdalin or nitriloside. It is found in only a few of the foods eaten in this country, and those foods are: millet, buckwheat, chick peas, and lima beans. It is also found in bitter almonds (not available here) but not in the ordinary sweet almond. It is not found in soybeans, but it is found in seeds such as apple seeds, peach kernels, and apricot kernels (from which it is extracted under the name of Laetrile).

Finally, I discovered that for a number of years, it has been possible to detect abnormalities in people who show no signs of cancer at present, but who will generally be found to have the disease as much as four or five years after the abnormality could have been detected. The accuracy of this simple, safe test has been shown to range from 80 to 90 percent.

If it is true that cancer is principally a vitamin deficiency disease, and that it can be detected (in the form of other symptoms) as much as five years ahead of time, then it should be possible to give such people a diet rich in Vitamin B-17, and find that the symptoms disappear and that the cancer does not, in fact, develop.

The ARF study of the cancer question began over a year ago when I was informed that a certain lawyer in this area had a considerable knowledge of certain psychic fields. On visiting him, I learned that his interest centered on unusual methods of healing, and that this began eight or nine years ago after his wife began to develop a tumor in her breast.

To begin with, she submitted to the usual surgical biopsy, in which the tumorous growth was removed. While she was still in the operating room, a "frozen section" test was made to tell if the growth was cancerous, so that if it were, the whole breast could be immediately removed. This test seemed to indicate that the growth was not malignant, so the breast was not removed, and because it was a Labor Day weekend, she was out of the hospital before it was discovered by the laboratory that the growth was indeed cancerous.

The couple were then told that the breast had to be removed at once, and the doctor was honest enough to admit that she only had a 40% chance of being saved from death by cancer within a few years.

Her husband did not want to believe this, and had another test made in the lab of a different hospital, but they came to the same diagnosis, and the operation was scheduled to take place in about twelve hours.

The woman was brave, and yet not certain that surgery was the right thing. A friend told her about a drug called Krebiozen, which she had also heard about previously, and they spent the next twelve hours looking for all the facts on this alternative treatment.

NEW RESEARCH INTO ASTROLOGICAL BIRTH CONTROL

WHILE doctors and psychiatrists argue over the harmful effects of abortion and the pill, while ministers declare the rights of the unborn child and legislators hassle about fine points of constitutional law, astrology makes a startling breakthrough into birth control!

In Philadelphia, Pennsylvania, the Aquarian Research Foundation has spent a great deal of time and effort to gather, test and publish the latest scientific data on natural methods of birth control, including the astrological method developed in eastern Europe, and for this purpose they have published computerized charts with lunar angles for every birthday since 1925. Their book, THE NATURAL BIRTH CONTROL BOOK, also gives a detailed account of primitive peoples who appear to use psychic control of conception as well as the experience of American women who are presently testing "mind over body" mental control of conception with great success. But the astrological method pioneered by a Czech psychiatrist is the main part of the book.

Interest was first aroused when Prentice-Hall published *Psychic Discoveries Behind the Iron Curtain* by Ostrander & Schroeder in 1970, and alerted new age scientists to the work of Dr. Eugen Jonas (pronounced Yo-nash) of Czechoslovakia. Dr. Jonas had struggled for more than a decade to get someone to listen seriously to his discoveries in medical astrology. As a young catholic psychiatrist, he lived and studied in Slovakia, close to the Hungarian border. When Hungary passed a law legalizing abortion, he wondered, "Couldn't there be a less traumatic way of going about birth control?" And so, in 1956, began his uphill fight to use astrology to solve the problem.

Dr. Jonas' first clue came when he noticed that women with a high nervous sensitivity seemed to go through a period of heightened sexual desire about once every 28 days regardless of their menstrual cycle. From his studies of both astrology and psychiatry, Dr. Jonas knew of the tremendous impact that the Full Moon had on human affairs: in many countries, policemen were put on extra duty because of the sharp rise in crime, full staffs are kept in mental hospitals to handle extra trouble, etc. He felt that there must be a relationship between lunar influence and his patients' recurring sexual drive, but what exactly was it?

An intensive search of all available data began. Dr. Jonas reviewed ancient manuscripts by great astrologers of the past; he gathered birth dates and histories of many women, charted facts and figures and discovered that an ancient Egyptian papyrus fragment linked the fertility of women to the phases of the moon. That was the key. He had found that, quite simply put, when the Sun and the Moon came to the same angle that they had at a woman's birth, she is likely to conceive.

Happily, Jonas put all of his information together and approached the Hungarian Academy of Sciences and the Czech Academy. He got their approval for more research, but no help at all with the means to do the work. More years would pass with Jonas working on his theory, refining and adding to the original concept. Eventually he came to believe that he could:

1) Ensure safe, reliable birth control without dangerous or undesireable side effects.

2) Help many seemingly sterile women become fertile.

3) Help women who have had nothing but miscarriages deliver full-term babies.

4) Help insure a healthier baby, eliminating some birth defects and mental retardation.

5) Allow parents to choose whether they will have a boy or a girl.

That these unique and amazing dis-

Finally, they decided to refuse the operation and instead went to a medical doctor in Lebanon, Pennsylvania. by the name of S. Allen Rutherford. This doctor told them that a niece of his was already undergoing treatment with Krebiozen for the same type of thing, apparently with success.

They also learned that ten years earlier, the Lankenau Hospital in Lower Merion, Pennsylvania, had supposedly run tests with Krebiozen on patients believed to be terminal. All of the patients showed some improvement. However, it was not possible to confirm this because, according to the rumor, when the report was sent in, the hospital was contacted by officials of the American Cancer Society, and told that if their report was not immediately rescinded, all grants for cancer research (which were considerable) would be cut off. Two doctors were supposed to have been involved with the test.

When my friends told their surgeon that they intended to go to Dr. Rutherford to receive treatment with Krebiozen instead of having the surgery, he was furious and refused to cooperate.

For about a year she received treatment from Dr. Rutherford. At first, more tumors developed, and then there was a standstill in the situation, so that they became quite concerned. Finally, they decided to go to Dr. Andrew C. Ivy in Chicago, who had been supplying the drug and the advice for its use. He then examined her and increased the dosage of Krebiozen by five hundred percent. After that, her condition began to improve and the tumors that had developed slowly disappeared without surgery.

She continued to make regular trips to Chicago for the next five or six years and all the symptoms of cancer gradually disappeared. At the same time, however, she followed the dietary advice of Dr. Rutherford and other friends, and stopped eating meat and foods that were not organically grown. She also stopped drinking fluorinated water. She is now quite healthy, requiring no Krebiozen or other treatment.

Dr. Ivy is a well-known figure in the medical world. He has been at the top levels of both the AMA and the American Cancer Society in the past and he was the representative of the United States government at the Nuremberg Trials after World War II. **Who's Who in America** gave him a column at least six inches long in Volume 32 (1962-1963). He gives Krebiozen

coveries did not produce world headlines suggests something less than openmindedness in scientific circles.

Finally, in 1960, Jonas was invited to the Bratislavia Clinic of Gynecology by Dr. Aurel Hudcovic. Here his fortunes turned for the better. People thronged to try his methods, and thousands of women were successfully charted. News of his discoveries filtered out and requests for more information began pouring in.

One of the people the discovery reached was Dr. Kurt Rechnitz, a university professor and former director of the Budapest Obstetric Clinic in Hungary. He plunged into an independent study paralleling Jonas' work and discovered that approximately 15% of the conceptions they were studying took place on or about the fifteenth day of the menstrual cycle regardless of the phases of the Moon. But he also found that the sex of the child was frequently determined by the sign the Moon was in at the time of conception. Additionally, he found that other pregnancies occurred as Jonas had calculated on or about the day of the mother's natal Moon phase. Dr. Rechnitz thus combined the rhythm method with Dr. Jonas' lunar theory. The combined method was used in volunteer studies made from 1965 until the invasion of Czechoslovakia by the Russians. Did that end Dr. Jonas' battle for recognition?

Alas, no, for it would be more years of frustration for our hero. Dr. Jonas' colleagues at home would brand him "insane" and went to great lengths to stop him from obtaining further degrees in psychiatry. But eventually, with other countries showing increasing interest and support, his own countrymen slowly came around. In 1968, the Czech government founded the Astra Research Center for Planned Parenthood. After more than a decade, Dr. Jonas had the facilities, staff, and even computers to carry on his work. Within a short period of time, the center was sending cosmograms to five continents at the rate of 1,000 per month. Within six months 9,000 applications had been processed, 2,300 of them for birth control. The Scientific Board reviewed 1,252 of these cases and found the method to be 97.7% effective. Although, theoretically, the method is 100% effective, in actual practice there were errors due to miscalculations and to difficulty in calculating rhythm fertility times by women whose menstrual periods were very irregular.

A year later, after the Soviet invasion of Czechoslovakia, the Astra center was closed, possibly because Dr. Jonas did not have much sympathy for the Russian takeover.

Research is now being done in the United States. Aquarian Research first published the details of the method in a preliminary edition in June, 1971. In November of 1972, they took a survey of 296 people who had bought the booklet and the results were as follows:

Over 125 people responded to the survey. Forty-five women reported that they were using the method themselves. Twenty-four said that they had been using the method over six months and twenty-one for less than that. They also indicated that twenty-two of their friends had been using the method successfully for various lengths of time. No pregnancies at all were reported by those whose use of the method was careful and complete — where they abstained entirely during the four day lunar fertility period and the whole of the rhythm fertility period. Six women said that they were using the method in order to conceive when they had previously not been able to do so. Five of these reported that they were successful. A new survey of people using the method is planned, and users are asked to send their name and address to Aquarian Research.

In May, 1973, the book was revised and enlarged and ten thousand copies were printed. These were sold out by April of 1974. In the meantime, a Dutch translation of the book has been published by Arcanum in Amsterdam. So much new information was received from readers and other sources, that an enlarged second edition was published this Spring. This edition contains more on psychic birth control and all the details of the ovulation method which shortens the time of abstinence and solves the problem of irregular menstrual cycles. The new book also discusses the latest research into the electronic detection of ovulation as a means of learning how the Astrological and Mental methods really work for different people. For women who are already troubled by a problem pregnancy, the book tells of spiritual communities where a mother may have her baby and leave it with a loving family until such time as she is able to care for it herself, all without charge.

In Bellingham, Washington, Ms. Nealy Gillette works with a woman's clinic where she has been setting up charts and teaching many women to use the new method with a high degree of success. A letter she received from Seattle is typical of the response.

"As a participant in astrological birth control for eight months, I find it not only working, but liberating. I feel my cycles as they come and can keep in touch with my body without relying on chemicals or devices to 'take care of things' for me."

Calculating one's approximate lunar fertility times only takes a minute. The user just looks up her birth date in the chart for her year to get the lunar angle for that date. If born in the morning, she adds six degrees as a safety factor. If born later in the day, she adds twelve degrees. She then looks at the chart for the current month (see actual ~~chart for 1974 with this article)~~ to see where in the month her angle occurs. Her lunar fertility period will be four days long. She is safe if her angle occurred over a day ago, or if she sees that it will not come for at least five days. If she is closer to her lunar fertility period, she would have to follow the

tested Jonas' work, has also come up with a theory that relates the influence that a man's and a woman's charts might have on each other in conception and sex determination. The theory has exciting possibilities but needs more research.

One part of Dr. Jonas' theory which also has not been tested here has to do with the health of the embryo. Jonas believes that certain interplanetary relationships have a negative effect on the fetus, especially for women who were born at Full Moon (lunar angle of 180 degrees) and some other angles. He believes that it is best for such women to avoid conception during their lunar fertility period and conceive during the rhythm or ovulation time when that occurs at some other time of the month. Women whose natal angle is less than ten degrees from Full Moon, 180, or close to 90, 270, and 360 degrees, should observe this precaution, Dr. Jonas believes. This is an area in which

Continued →

without charge, accepting only donations.

WHY IS KREBIOZEN NOT TESTED? It seems strange that the Cancer Society and the other government research programs have not made thorough tests with Krebiozen as a control or cure for cancer. Why is this? Some hints are found in the Congressional Record of the Senate on page 22493, where we find the following statement of Senator Paul Douglas on December 6, 1963:

"In connection with innuendos of fraud, it is significant that two of the most prominent drug companies in the United States have each offered approximately $2,000,000 in cash and royalties for the 'exclusive right throughout the world to manufacture and sell Krebiozen.' (photostats, exhibit 13)

"It may be assumed that these firms arrived at a careful judgement of the intrinsic worth and potential of Krebiozen before they made these extraordinary offers in writing.

"Dr. Durovic (the discoverer of the drug) refused these offers mainly because he understood that the companies planned to retail the drug at a price of $30 to $40 per ampule instead of the $10 which he thought was right." (Illinois Legislative hearing, p. 1139)

Dr. Durovic then went to Dr. Ivy and showed him the discovery. Dr. Ivy tested Krebiozen and accepted it and has been harassed almost from that time to this. Over four thousand people have been given Krebiozen by Dr. Ivy. 42% of them have experienced regressions of the cancer, and 86% received help in some way. Yet 96% of those people were looked upon as "terminal cases"; that is, people for whom medical science held no hope."

instructions exactly to be safe. Of course, to be as safe as possible she must also avoid relations during her ovulation time or rhythm fertility period. The fact that a woman can become pregnant at times other than the fertile time between the menses, sometimes even during the menses, was little known until Dr. Jonas made his discoveries. For women who have gone off the pill, astrological birth control is not to be used without mechanical devices until the menstrual cycle is normal again. This can vary with different women from a month to over a year.

SEX DETERMINATION

According to Dr. Jonas, the sex of the offspring is usually determined by the sign the Moon is in during the lunar fertility period in which the child is conceived. The Moon's position at the hour the natal angle returns is what matters. This is most exact when conception occurs within a lunar fertility period and outside of the rhythm fertility period. The Moon enters a new sign roughly every three days. If the Moon is in Aries, Gemini, Leo, Libra, Sagittarius or Aquarius a boy will tend to result. Otherwise, the baby is likely to be a girl. If the Moon is close to the cusp between signs when the natal angle occurs, the sex is not predictable. An accuracy for this method of 98% is claimed. Computerized charts and exact calculations for sex determination are given in the book but the 98% figure has not been checked in this country. Dr. Barna Balough of Hungary, who observed and further research could easily be done. The Aquarian Research Foundation would be happy to hear of anyone who is able to study the matter further in a statiscal way.

Another area of interest is the relationship of the planetary patterns at the time of conception to the patterns at the time of birth. "It is Jonas' view that. . .a fetus that is developing in the mother's body cannot normally be be born at just any time, but only when the effects of the universe are similar to those at the time of conception. . .The initial impulse is the most important as it will shape the patterns of existence, health, personality, and favorable periods." These patterns are discussed in the book, *Body Time*, by Gay Gaer Luce. In it, she puts together an immense amount of information on biological rhythms in humans. This is one of the most important new areas of medical research.

Since 1971, about 20,000 copies of THE NATURAL BIRTH CONTROL BOOK have been sold, and the Foundation receives many letters from readers. To our amazement, we have yet to receive a report from a user telling us she has used the method fully and carefully and became pregnant unintentionally. We can't believe the method has never failed, but if it had failed often, we would have had at least a few letters telling us of the failures. So, we think astrological birth control is working extremely well, and that makes us wonder why the world is so slow to accept it. In the last few months, we have had several orders for the ~~book from Planned~~ Parenthood groups and one from the Dept. of Health, Education and Welfare, but it seems unfortunate that the method isn't better known.

People's unwillingness to accept new ideas has been a plague to mankind ever since the days when the world was flat and people were burned for saying it wasn't. People tend to resist change even if it's for everyone's good in the long run. Insecurity, fear and ignorance make us turn away from the unknown instead of looking forward.

The Aquarian Research Foundation seeks ways of helping overcome society's resistance to change that prevents in large part the new age of Aquarius from becoming real today.

People ask us, "How do you help people's minds to open?" We say that when they have remarkable new experiences, such as a natural birth control method that solves a difficult personal problem, then the mind becomes more open. A person may then take a second look at what seemed to be "ridiculous nonsense."

IS FDA GUILTY OF FRAUD? In a paper which was delivered to the Congress on Health Monopoly at the New York Hilton Hotel, April 7, 1965, Dr. George W. Crane, M.D., Ph.D., said the following:

"As a former five-year research scientist with the National Research Council and the Carnegie Institution of Washington, D.C., I charge the Food and Drug Administration with fraud! The FDA has repeatedly claimed in print that Krebiozen is nothing but Creatine. It says they are **identical!**

"Yet, in Senator Paul H. Douglas' office in Washington, D.C., on November 13, 1963, I saw the chemists' report from the FDA, saying that Krebiozen is brown in color but Creatine is **always** white! That one chemical difference is enough to demolish the entire fraudulent charge of the FDA. But also note Dr. Anderson's 29 valid differences in their spectrographs, plus Professor Roy Whistler's 16 different chemical substances in Krebiozen which are not in Creatine at all! The FDA knows these facts but deliberately chooses to ignore them in its mania to smear Dr. Ivy and Dr. Durovic.

"The FDA has released a news story claiming that Krebiozen has no clinical value in treating cancer. Yet the FDA has never used Krebiozen on mice, rats, or human cancer patients.

"Since 3,000,000 cancer victims have died in the 14 years of this controversy, many of whom begged to be treated with Krebiozen, yet were denied such treatment on the grounds that Krebiozen was a 'quack' remedy, totally ineffective, suppose that a 'double blind' test should now be made and Krebiozen were proved to be helpful in only one case out of every three! That would mean one million cancer victims have died needlessly and their murder would be laid at the feet of the FDA. What would be the outcry from all the relatives of those needlessly dead victims of FDA bungling and deceit? Is it any wonder that the FDA constantly places roadblocks in front of **any** suggested clinical test?"

LAETRILE ALSO BANNED BY FDA Today one hears more about another drug for cancer which is known as Laetrile. Recently, a major TV network did a documentary on the subject of Laetrile therapy for cancer, and they showed a hospital and a clinic in Tijuana, Mexico, in which people who were being treated with this medicine appeared to be quite well, though they were said to be "terminal cases."

During the documentary, the program switched back several times to a doctor in New York City who said that Laetrile was not an effective cancer remedy; that they had tried it and it does not work.

Dr. E. T. Krebs is the son of the doctor who first used Laetrile as a control for cancer. Dr. Krebs works at the John Beard Memorial Foundation in San Francisco, and I was able to speak with him at some length on two occasions.

Briefly, he told me the following: Laetrile is a natural food substance which is obtained from the kernels of apricot seeds. It is now believed to be a vitamin of the B complex, and since it is seventeenth in line of discovery, it is referred to as vitamin B-17. Dr. Krebs told me that he is convinced that cancer is chiefly a vitamin deficiency disease, due mainly to a deficiency of vitamin B-17, which is not found in most of the foods common to the diet of "civilized" people. He named those foods which are mentioned at the start of this paper.

He told me that the extraction of this substance is carried on in Mexico and some other countries, including Germany, where it is supplied by Dr. Hans Nieper in Hanover. The vitamin can be given orally in quantities up to about one gram every eight hours. Larger doses can be given by injection but not eaten because they tend to upset the stomach. He said that if a person has pain from a cancer, then it would usually be possible to overcome the pain within 24 hours by a dose of B-17 amounting to about one gram each eight hours. That is one way of telling if the medicine is effective.

He also told me that I could obtain Laetrile under the names of amygdalin, nitriloside, or vitamin B-17 from a biochemical company in Ohio, and that the price there was $2.90 per gram. This company sells chemicals to the research drug industry, and they will not sell it if they are told that it is to be used for the treatment of cancer because the FDA considers it an illegal drug when used for that purpose. It is, however, not poisonous, and it does not get you "high."

Briefly, the way in which the chemical is thought to act is that it contains the compound cyanide, which is a strong poison. However, it is contained in such a form that it has no harmful effect on normal health cell tissues, but when it comes into contact with cancerous cells (which function in a different way), the cyanide is released and kills those cells. Dr. Krebs believes that if a person obtains sufficient quantities of vitamin B-17 in his diet, he will never get cancer.

After speaking with Dr. Krebs, I left San Francisco and proceeded to Tijuana, Mexico, where I visited both the clinic and hospital, which is operated by Dr. Ernesto Contreras and his son, who is also a doctor. There, I spoke briefly with some of the outpatients who had been treated and were now able to leave the hospital, coming only when necessary to obtain a supply of the vitamin or receive an injection. They seemed to be cheerful and not in pain at all.

Then I had a long interview with Dr. Madrid, who intended to go on to become a neurosurgeon. He had been working with Dr. Contreras for eight months, and he permitted me to tape record the interview. He told me that almost all of the patients who had been coming to them until now had been advanced, or "terminal" cases. But that recently, for the first time, they were getting some patients who had only recently been diagnosed as having cancer, and had come to them before any surgery or radiation treatments had been given.

Therefore, he believed that in the next few months it should be possible to see if complete cures could be brought about. Otherwise, he was only prepared to say that Laetrile is a control for cancer which takes away the pain and eliminates the need for narcotics, but that the patients that had been coming generally had such advanced destruction of organs that they were not likely to live very long.

However, from what I had already seen at the clinic, I had the feeling that he was being very conservative in speaking with me and did not want to say that they were "curing" cancer for fear of trouble. The outpatients that I met at the clinic seemed so well that I could hardly have taken them for "terminal" cases. Both Dr. Krebs and Dr. Madrid had spoken very highly of a recent article in the magazine **Prevention**. It is in the December, 1971 issue, and the magazine can be obtained from **Prevention** magazine, Emmaus, Pennsylvania 18049.

According to that article, the FDA forbids the sale of bitter almonds (which are eaten commonly in China) because they contain cyanide, and the FDA thinks that they might kill you. On the other hand, some primitive people eat up to one half gram of cyanide a day in natural foods without ill effect. According to Dr. Krebs, you would have to eat about 20 teaspoons of ground apricot kernels to get that much.

Dr. Madrid said that most of their patients come in with severe pain or drugged with narcotics, but after treatment with Laetrile, they begin to feel better; their mental attitude improves, appetite increases, and their general condition begins to get better. They do not have a complete recovery, but they live a more or less normal life without fear of pain or immediate death.

Laetrile, they believe, arrests the tumor process, but does not make the tumor disappear. However, for patients with early cancer and no previous treatment, one to three grams of Laetrile a day may be sufficient to cure it. Where the tumor is readily accessible, it is removed by surgery while the patient is taking the drug.

A diet which is fat-free and low in animal proteins is recommended. Little meat, but fish is O.K. Dr. Madrid thought that foods containing B-17 might help a person with a small tumor, but that once the tumor process has begun, high doses of medication should really be given to try to arrest the tumor. But such foods should be eaten by all people as a cancer preventative.

PREDICTION OF CANCER After I returned to Philadelphia, I learned about another very important discovery, which is the main reason for this article. I was given a paper entitled "Observations on Disturbances in Neuromuscular Coordination in Patients with Malignant Disease" by Alfred Kanfer and Daniel F. Casten, M.D. This paper was originally published in the Bulletin of the Hospital for Joint Diseases (New York City), Volume XIX, No. 1, April 1958. It describes Mr. Kanfer's discovery that microscopic examination of handwriting samples can indicate whether or not a person is likely to have cancer at present or at any time in the next five years.

Mr. Kanfer has been able to detect and analyze a change in handwriting as much as five years before a malignancy was found to be medically apparent. The changes were visible in microphotographs using a magnification of ten to sixty times. The method of detection could be taught to other people. The changes which are manifested in these microphotographs are peculiar to cancer, and are not brought about by any other conditions.

When the paper was written it was based on a study lasting twelve years and covering thousands of cases. The research was supported in part by a grant from the American Cancer Society.

After reading the paper, I was very much impressed and spoke about it with Barry, the friend who had supplied the paper. He also managed to locate the present place of employment of the author, Alfred Kanfer, who is working at the Strang Clinic for preventive medicine in New York. I then went down to New York City and visited Mr. Kanfer at his place of work in the basement of the clinic.

Before going to see him, I came to an interesting idea: if it is possible to determine years ahead of time that a person is going to get cancer, then might it not also be possible to change that trend by having that person eat a diet which is rich in vitamin B-17? If cancer is due to a deficiency of that vitamin, then it should be possible to prove that by observing that the patient now does not develop cancer, and that his handwriting in time reverts to normal. If this were to happen, it would prove both these fairly well.

The handwriting test itself is a simple one, and Mr. Kanfer informed me that he does about sixty of them each day. He said that a handwriting sample is taken from each of the patients that are examined by the clinic. He said that the accuracy of his test is found to be between 80 and 90 percent.

The Strange Clinic, also known as the Preventive Medicine Institute, gives complete physical exams for the prevention and early detection of all kinds of diseases. Their brochure gives a list of the various types of tests that the patient is given, but does not mention any test based on analysis of handwriting. While I was with Mr. Kanfer, I asked him why this test was not generally accepted by the medical profession and his answer was typical: "If I would use a drop of blood in my analysis, they would accept it, but because it smacks of 'graphology,' it is not considered to be 'scientific.' "

Mr. Kanfer offered to help with the experiment that I am suggesting, and also offered to permit a wider publication of an article about his work, provided that he would be permitted to see and check the accuracy of the article before it is published. He told me that when the publisher agrees to publish something, he will send the necessary photos and information.

The fact that the Strang Clinic has employed Mr. Kanfer for six or eight years, and that they supply him with handwriting samples of each patient, seems to indicate that the published study is as accurate as it is claimed to be by Dr. Casten and the others who worked with him.

Another possibility which may be considered is the measurement with an electromyograph of the electrical currents in the nerves and muscles of the fingers which determine the variations in the character of handwriting. I will be in touch with friends who have access to such equipment, and see if they are interested in setting up the experiment. It may be possible in that way to improve on Mr. Kanfer's discoveries and bring them into the realm of "accepted technique" at the same time.

If cancer is indeed due to a B vitamin deficiency, and if that same deficiency can cause the change in neuromuscular coordination on which Mr. Kanfer's test is based, it could also affect the brain in other ways. While in California, I also asked Dr. Krebs about the coincidence of cancer with certain mental states, and he confirmed that there was such a relationship, but did not think of the mental state, as the causative factor. However, it is interesting to note that Dr. Madrid made special mention of the fact that a patient's mental attitude improves when he starts the Laetrile treatment. This may be more than a coincidence.

I also asked Dr. Krebs what he thought about Krebiozen, and he told me that the problem with Krebiozen is that it is very difficult to be sure if you have the real stuff or not, whereas the analysis of amygdalin is relatively simple.

As many of you know, the Aquarian Research Foundation is not, in general, dedicated to medical research, nor is it our purpose to attack or expose fraudulent activities on the part of governments or other people.

Our work is to find ways of overcoming society's resistance to change so as to help a beautiful New Age to come quickly and peacefully. But to do that we have to open to the Spirit and act upon whatever insights are given to us, even if it is only in the production of a humble trap for cockroaches. It is my hope that other people, better qualified than myself in the medical field, will follow up on some of the discoveries told about above, and I would be happy to help in any way I can.

NATURAL MEDICINE—VITAMIN C, ETC. Since publishing the information on natural cures for infections using vitamin C, garlic and vinegar, I have received a small book telling about lots of research on the use of vitamin C and other vitamins as curative agents in many conditions. It is written by Fred Klenner, M.D. The work seems to be scientific enough, but the layout of the book leaves much to be desired. Still, it is **well** worth having. Send $2.00 to Graphic Arts Research Foundation, 114 W. Kinzie Street, Chicago, Illinois 60610. Ask for **Key to Good Health, Vitamin C.** It is sold with money back guarantee.

TRIP TO ARIZONA: The Universal Party had their convention in Flagstaff, Arizona, April 8 through 10, 1972. They nominated Gabriel Green for President and Daniel Fry, Ph.D. for Vice President of the United States. Green is president of the Flying Saucer Clubs of America, and Fry is the author of **The White Sands Incident**, in which he claimed to have ridden in a saucer.

The Universal Party invited me to attend the convention at their expense, so I was able to fly there and back. On the way, I stayed with Tim Zell of the Church of All Worlds, and as he was also invited to the convention, we were able to go there together. Unfortunately, I have not yet been able to verify the claims of Dr. Fry, and so I have not been able to put time into the work of that group. They do, however, express the idea of a "New Age" party which would seek to end war on this planet and seek unity with people on other planets.

SILVA'S METHOD IN THE WHITE HOUSE: While in Arizona, I realized that a presidential candidate who was sincerely concerned for the good of mankind would have an amazingly good chance of getting elected if he was also a trained psychic or a graduate of the Silva Method.

You may remember that 500,000 or more Americans wrote letters to the North Vietnamese delegation in Paris asking that war prisoners be freed. Politically that may have been very naïve, as prisoners are never released till a war ends. But few people realized that this flood of letters was started by the action of a tiny group of Mind Control graduates in Texas, doing what seemed to be a loving action with psychic methods.

This would indicate that a sincere candidate who was a Mind Control graduate and who had graduates working for his campaign, might have a tremendous advantage—especially if thousands of other trained psychics were with him spiritually. Therefore, it might be a very worthwhile project to try to get the most honorable of the candidates to take the Silva experience, which would also be of tremendous help in bringing the world toward a New Age in a peaceful way, so to have a president with that understanding would not be a bad thing.

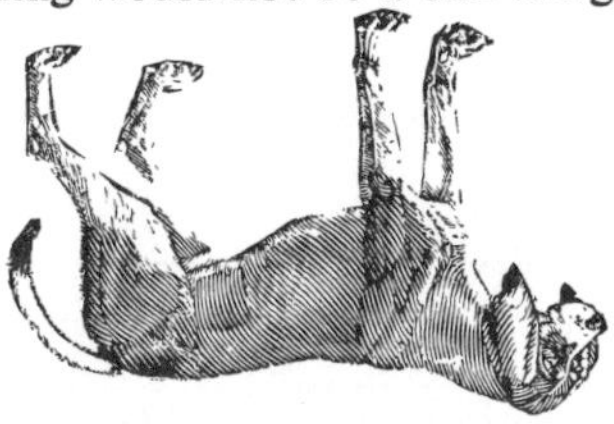

LADY HAS TEN PUPPIES AND A KITTEN! Ten pups were born to Lady on April 14, but the kitten came from the SPCA a week or two before. The kitten already catches mice, but the pups have not opened their eyes yet. Both Lady and her mate (who is Belgian shepherd) are very gentle, but also very good watch dogs.

ANOTHER TRIP TO CALIFORNIA

The purpose of our second California trip in December, 1971 was only partically clear to me in the beginning. While there during the summer, I felt very much hindered in not having a car and not being able to go see all the people and groups that I felt important.

Additionally, through the **International Cooperation Directory** (17819 Roscoe Boulevard, Northridge, California 91324), I discovered that there were many other groups that I had not heard of who could be important to the work we are doing, as well as the coming New Age.

NEW AGE GROUPS: In fact, so many of these groups were concerned and aware of the coming New Age that it amazed me, and I felt that I had to learn more.

In the back of my mind, there was also a concern to visit the hospital in Mexico where cancer patients were being treated with Laetrile.

CANCER CAN BE CONTROLLED

I now realize that one of the most important reasons for the whole journey at this time was to learn about the control—and possible cure and prevention—of cancer.

The work of the Aquarian Research Foundation has never been predetermined in the kind of logical way that most institutions are operated. Our plan has been to use intuitive ability (or openness to the leading of the Spirit) as a means of deciding what to do.

For that reason, we did not calculate whether the trip was worth the time or money it would have cost, but, having been given a suitable vehicle **and** a feeling that a second visit was in order (though not very logical), the increased awareness of the Silva Method (going to an Alpha brain wave state and getting answers intuitively) was used to make this decision. I have used this method frequently, and have found that the results usually prove to be much better than expected.

On arriving in Los Angeles, I realized that I did not even know where the Mexican cancer hospital was, and so I had to call Philadelphia to find out. I was then told that Pat Wolfe of the Scientific Research and Civil Rights Foundation was in Los Angeles.

SCIENTIFIC RESEARCH & CIVIL RIGHTS FOUNDATION

This organization is a group that works through the courts in order to assist medical research workers who have made new discoveries in the field of medicine and healing, but who are too far ahead of their time to be accepted by the current medical establishment. Had they been around at the time of Wilhelm Reich, they might have saved his life.

Pat Wolfe told me a great deal about himself and his experiences, and he gave me the address and phone number of Dr. E.T. Krebs, who is the son of the doctor who discovered the drug Laetrile as a control for cancer. I was able to have several long talks with Dr. Krebs when I got to San Francisco.

Karl Heussenstamm, who has been a member of the board of directors of the International Cooperation Council (they publish the directory) for a long time, invited me over for an evening, and we talked at length about his ideas and mine. He is proposing to start a community for people who are seeking to bring about a new lifestyle in conformity with the age to come.

Afterwards, I went to the southern end of Los Angeles and visited with John L. Davis, president emeritus of Chapman College, and the man who started me on this whole adventure back in 1947 when, as the dean of Hiram College in Ohio, suggested to me that a years leave to live in a Paraguayan community (the Society of Brothers) could be worth more than a year in any college. As you know, I spent over 15 years with the Society of Brothers—the Bruderhof—in South America, and then in New York and Connecticut when the Society left Paraguay for the Northeastern United States. It was really nice to spend an evening with Dr. Davis after so many years.

On Sunday, I attended a meeting in Northridge of the International Cooperation Council, and met its director, Leland Stewart. The meetings are held at his home, which is a center for I.C.C. activities.

ABILITISM was the thing most discussed between myself and Craig Lesser, a mind control graduate whom I met after the I.C.C. meeting at Leland Stewart's home. Abilitism is a program which claims to straighten people out emotionally by group interaction. I believe that Abilitism is an offshoot of Scientology (but I'm not sure), and is the brainchild of Charles Brenner. I would have liked to have had the time to meet Brenner and experience the work he is doing, or at least meet more of the people who have experienced Abilitism. From the report I had, I felt that Brenner was into a rather mystical trip, and so far, I am skeptical, but his main emphasis seems to be on teaching people to relate more deeply to one another, and that seems good. I don't have an address for Brenner's Institute of Ability (which Craig said was a commune), but I am certain that you can get it

from the International Cooperation Council in Northridge.

Craig also gave me directions to a place I had heard of before, called "The College of Universal Wisdom," in the desert near Indio, California. They publish a newspaper that has a lot to do with flying saucers, so we decided to visit there on the way home. Unfortunately, we did not find it as scientific as Craig seems to have thought it to be, and, in fact, we have a strong suspicion that it's quite a swindle.

JOURNEYS OUT OF THE BODY!! is a book by Robert Monroe. He is an engineer and a pilot who has had experiences with what is sometimes called "astral travel" for about 12 years. In our last newsletter, I mentioned that book. Now, while in California, I went to visit Charles Tart, who wrote the introduction to Monroe's book. Charles is a younger man that I expected, but he does know a lot about altered states of consciousness and has edited a book by that title. He has made laboratory tests of Monroe's ability to travel on the astral plane, and he told me that the book is a true and scientific account of astral travel experience. He also saw some of the same small errors in the work that we had noticed, but he did not think that Monroe was exaggerating his experiences. **Journeys Out of the Body** is well worth reading.

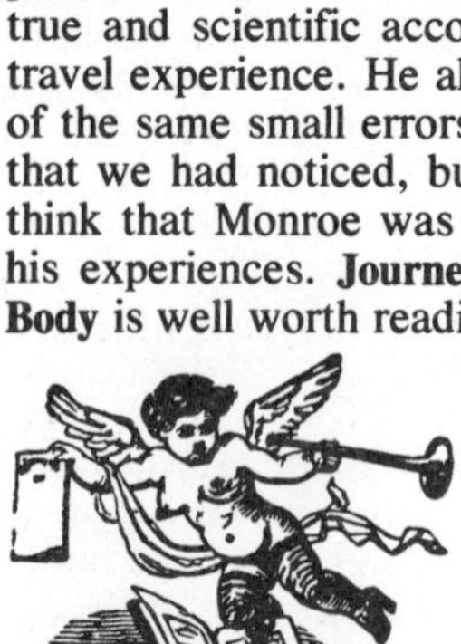

SEX AND THE SOUL: Thoughtful readers of Bishop James Pike's book **The Other Side** will remember that Pike asked of his son, "Do you think of people there as male and female," and, "Is there something like—intimate expression?" And the son replied: "Sex? Yes, there is sex. But it is not like it is there. It is not physical. Of course, but actually there is less limitation. It is more obviously like what sex really means. Here you can actually enter the whole person. It is like you are in fact merging —becoming one."

This report seems to fit very well with the experience on the spirit level which was frequently given to Monroe in his **Journeys Out of the Body**. Monroe is honest and personal enough to describe some of his out of the body relationships with other people, and thereby he gives information on matters of great importance. It was good to hear from Charles Tart that the book seemed genuine.

We also spoke with Tart regarding dream control, and he said that Joe Hart of the University of California at Irvine was the best one to teach about that. Tart knew of no one who could teach astral travel.

BEAR TRIBE: We then went to visit a group of the Bear Tribe, then under the guidance of Tommy Gunn, who was the first Bear Tribe member that I had met. We found Tommy at home in a very comfortable sort of quonset hut, which was part of a larger farm leased to them for very little from a friend. He told us that about a dozen of them lived there. He also told us of several other winter hideouts of Bear Tribe people, and of Sun Bear's whereabouts in Reno, Nevada. My impression was that of a very well ordered small community by themselves for the winter, waiting for the summer before new people would be expected.

The address of Tommy Gunn is Box 109, Route #1, Clarksburg, California.

We then went to see Shiela Daar, editor of **Ramparts** magazine, and we discovered that **Ramparts** was planning to publish an article on cancer research originating from England. Then we went to see Jim Baird of the American Friends Service Committee, and found the AFSC working on a project to help people find new ways to earn a living. They gave us a copy of their book **Working Loose**, which asks that aside from making money what are our chief desires in life? It then goes on telling how other people have managed to find new locations they really enjoy, and it gives them some hints about how you might do the same. Sells for $1.95. Write AFSC, 2160 Lake Street, San Francisco, California 94121. The book is important because working at a job you don't believe in is a major cause of death in America. I mean **real** death, not just the loss of the physical body. Deeper insights come to people when they are doing the things they believe in.

We then went to see John Bergerm, who is in charge of the Alternative Features Service (AFS), and that turned out to be an important contact. He gave us the name of their science editor, Saul Paul Sirag. Saul understands science: medicine, chemistry, and physics are all his strong points. He also seemed to understand what ARF is about, and we spent a fair amount of time together.

TAPE RECORDING SPIRIT VOICES Recently, in England, there were two books published which deal in a scientific way with the direct recording on magnetic tape of the voices of people who are no longer living in a physical body.

The first to publish widely in this field was Dr. Konstantine Raudive, a German psychologist. When Colin Smythe, Ltd., was asked to publish a translation of his book under the title of **Breakthrough**, they were very interested, but wanted some scientific verification in England of the reality of such spirit voice recordings.

Dr. Raudive was asked to come to England and demonstrate such recordings and the method by which they were made in the presence of some of the best British professional recording engineers. Then the doctor was asked to be present only while the engineers themselves attempted to duplicate the experiments, and this was done in shielded rooms so that no radio station interference could be responsible for the voices recorded.

Afterwards, these and other engineers were able to produce such spirit voice recordings without the presence of Dr. Raudive. A large amount of publicity was given to the whole matter both in the press and on

TV, and many church people were also required to take a stand on it. The controversy and the stands of various people, and the testing that was carried on before **Breakthrough** was published, was described in the second book called **Voices from the Tapes** by Peter Bander, published by Drake Publishers in New York City. The technical details of spirit recording, and the theories which Dr. Raudive espoused to explain it, are found in **Breakthrough**. I must say it is strange that so much publicity can be given to the matter in England and hardly a word in the American press. (I apologize: an article about this appeared in **Fate** in 1970, and another in the **National Enquirer** in early 1972.)

TAPE RECORDINGS FROM BEYOND —DOING IT YOURSELF!

Here are the full instructions for the tape recordings of spirit voices. You may reprint the article without permission.

We have now to report a breakthrough into the realm of science. Thousands of experiments are producing amazing evidence that people who have died are able to record their voices directly on magnetic tape running through an ordinary tape recorder!

It would not be unreasonable to write about this discovery in the most sensational manner, with pictures of the experimenters and descriptions of the most amazing voices that have been recorded, but the real and convincing proof is to let you record such voices yourself. Here is the information you need to do that!

About 1964, the Swedish author Friedrich Jurgenson placed a tape recorder in the fields in order to record the songs of birds and sounds of nature. On listening to the tape, he noticed that Rufus, his Great Dane dog, seemed to take a special interest in a portion of the tape where there was nothing special to hear. On closer inspection, a weak but intelligible human-like voice could be discerned at a level just above the background noise of the tape! Later, similar voices were heard, and they all had a rhythmic quality quite unlike anything which might have been produced in the normal manner. With a little practice, almost anyone was able to hear and interpret some of those voices, so there was no possibility of their being an illusion of any kind.

Mr. Jurgenson discussed this phenomena with Konstantine Raudive, a psychologist friend who was living in Germany at the time, and Dr. Raudive came to listen to the tapes. That was the beginning of Dr. Raudive's years of experimenting with the "voice phenomena," which we now have good reason to believe are the voices of people who have passed out of the "physical dimension." Dr. Raudive concurs in this belief, as do many others.

The voices are different from ordinary speech in that they speak about twice as fast as a human voice and have a strange rhythm about them. They rarely occur in ordinary recording work, but are more frequently heard when they are invited to speak by the person who is doing the tape recording, and they sometimes call his name or volunteer their own.

Though Dr. Raudive and others who have worked with him have recorded over 72,000 voices, none of them seem to be able to carry on a lengthy or sustained conversation. It seems as if the invisible personalities require a considerable time to build up the energy for a short word or phrase, and are then not able to speak again for some time. Scientists who are now studying this phenomenon are attempting to design recording equipment which might possibly be able to supply additional energy, so that the entities could carry on at greater length with less effort, but little is known of the manner in which they record their voices, so the research is still a shot in the dark.

Dr. Raudive published a book about his work in 1968, and later gave a copy to the English publisher, Colin Smythe. Smythe's associates read the work and were inclined to disbelieve, except for one thing; there were the names of a considerable number of well-known German scientists who claimed to have worked with Dr. Raudive, and to have heard and recorded such voices themselves. Upon investigation, the publisher learned that these reports were correct.

Peter Bander, one of Mr. Smythe's associates, tells of his own experiment which convinced him that the voices were real. It was actually Smythe himself who attempted to try out one of the recording methods which Raudive described. He told Peter that he had heard something which might be such a voice just before leaving on a trip to Ireland.

Mr. Bander tells the following in his excellent book called **Voices from the Tapes:**

"... I found a tape on the recorder, and next to it a paper with the number 126 written on it. I rewound the tape to 120 (on the counter) and played it back. I heard nothing. I played it back five times, and was just about to give up when I suddenly heard this rhythm. I played it again and once more. It is very difficult for me to narrate clearly what happened next; suddenly I heard a voice—quite distant, but very clear. I played that part of the tape again because I was absolutely certain that I had fallen victim to my own imagination. The voice was—if anything —clearer. A woman's voice said in German: 'Mach die Tur mal auf' (Why don't you open the door). This in itself might not appear significant to an outsider . . . but for two facts. As soon as I heard the voice I recognized the speaker; although the voice spoke terribly fast and in a strange rhythm, I had heard it many times before. For eleven years before her death I had conducted my entire correspondence with my mother by tape, and I would recognize her voice anywhere. And this was my mother's voice. Secondly, Colin Smythe cannot speak German and in any event is quite incapable of playing a deceitful trick of this kind on anybody.

"Later that evening I went back to the tape recorder and played the tape over and over again. There was no doubt about the voice being on the tape. I no longer had any difficulty hearing it right away. For somehow my ear appeared to have tuned in to this unusual frequency. I tried hard to find some rational explanation and on Monday, without going into details, I asked two of our junior assistants in the firm to listen to the tape. They listened with stereophonic earphones ten or twelve times, then wrote down what they heard phonetically, and although these young men are not trained in phonetics it soon became clear that they were hearing the same thing that I had heard. On Colin Smythe's return from Ireland the following day, I asked him to listen carefully to his own recording, and after playing it five or six times, he noticed a voice but was still unable to identify language or contents."

Later, Peter Bander wrote a book about this and other similar experiments including some with England's best professional recording engineers. As a result, the firm published Konstantine Raudive's book under the title **Breakthrough** and Bander's book entitled **Voices from the Tapes** subtitled **How Dead are the Voices? Breakthrough** (391 pages) is $10.00 and **Voices from the Tapes** is $6.00. We highly recommend both books to our readers. In addition, there is also a phonograph record (33rpm) on which Dr. Raudive has recorded selected voices. The record is available for $3.00 from Vista Productions Recording, 64A Landsdowne Road, London W11 2LR England. The record might help the experimenter learn what to look for.

Richard Sheargold, who is a researcher for the British Society for Psychical Research, had previously been a radio amateur for 40 years. He tried to repeat the experiments, but was unsuccessful until he began to use earphones instead of listening to the speaker of the tape recorder. After his success, he published a newspaper article about it and then wrote a paper called "Hints on Receiving the Voice Phenomenon" to assist other experimenters. Since then, he has been swamped with requests for that information, and has asked me not to reprint his address. But the whole of this paper is reprinted below. The notes in parenthesis are mine; they may help because I am also an electronics technician.

PART II

"Hints on Receiving the Voice Phenomenon", by Richard K. Sheargold, England, 1972.

GENERAL: The voices you are trying to receive are **very weak**. This cannot be overstressed. Although there is always the possibility that you may be favored by a voice much louder than usual right away, this is greatly the exception and you must expect to exercise patience and perseverance. It may be any period between a few days and a few months before you hear anything at all. You should devote a short period each day to listening. Half an hour should be ample, as long sessions will only be tiring. Follow the hints given below closely as they have yielded results.

The use of headphones with earmuffs is practically essential. (These are available for about $6.00 and up at electronics stores.) Suitable phones will probably be the 8 ohm stereo type but a guide to this is probably given in the literature you received with your tape recorder. If not, any dealer could advise you if you tell him the type of recorder you have. Not only does the use of headphones assist your concentration, but it helps to maintain family accord! However, since I have found that there are certain people unaccustomed to earphones who seem to hear better from the loudspeaker, it is suggested that until experience with the earphones has been gained, any recording is to be played back at least once on the speaker. But you may be definitely assured that the weaker voices which are usually the most informative will not be audible on the speaker.

One word of caution is necessary. The noise made by the tape passing through the head plus that generated by the amplifier itself is that of a loud rushing sound akin to that known as "white noise" by engineers. Prolonged listening to this produces a very well-known effect when voices tend to be heard which do not really exist. It is this effect which is used by the natives with their "conch shells" (or children with sea shells). These spurious voices are usually male and of course words cannot be deciphered. I first encountered this phenomenon years before the war during my early experiments using super-regenerative receivers which produce a continuous hiss. It is easily possible to distinguish these spurious voices from the real phenomena we are seeking because obviously the spurious voices cannot be repeated on playback, whereas the real voices, no matter how faint, can be repeated as often as is desired. For your peace of mind I may add that after the experience has been gained and the real voices heard, you will in all probability cease to be troubled by this "conch shell" effect at all. I very rarely hear it myself now. (I would suggest that new tape always be used for these experiments, and that it be of the good quality, low noise variety. Also, the tape head should be clean, free of magnetism, and not too worn. If a considerable degree of hum is a problem, and the set is a portable unit, then the recording could be made out of doors away from power lines, perhaps in a parked car, and there will be no hum.)

METHODS OF RECORDING: Various methods of recording are described by Dr. Raudive in **Breakthrough** and illustrated on his record. Personally, I have at present only used the microphone and diode methods. The microphone method simply means using the tape recorder with a microphone in the usual way. The diode means a very simple electron device which is in effect a "crystal set" with a modern diode as a crystal and an R.F. Choke as a tuned circuit which is shunted by a 100,000 ohm resistor and coupled to the input of a tape recorder. (See circuit details in diagram). Voices received via the diode tend to be rather weaker than those via the microphone, but they seem rather more natural in sound and the phrases they speak are longer.

MAKING A RECORDING: In order to record the voices by any method it is almost essential to record at maximum level. Further, since we are going to listen most intently to the tape on playback, it is essential to avoid any deafening noises appearing on the tape. So the correct method as far as I am concerned is as follows:

1. Microphone: Place the microphone some distance away from your recorder in order to avoid picking up the sound of its motor. Make your preliminary announcement (see later) with the gain just high enough to put a reasonably loud voice on the tape. Immediately after the announcement is finished and you desire to record any voices, turn the gain control up full and sit **perfectly still**. Remember that with the gain up full, even the laying down of a pencil on the table

can produce a defening crash on the tape!

(Mr. Sheargold must have had a pretty sensitive recorder for that to be so—perhaps such sensitive instruments with lots of gain are also required. But of course, with intense careful listening it doesn't take much on an almost blank tape to produce what seems like a deafening crash. Perhaps a more sophisticated arrangement would be to have the microphone a good distance from the machine so that when the operator speaks in a low voice it just records well even with the gain set at full because of the great distance between the operator and the microphone. That way, there would be no need to touch the gain control during the procedure. Perhaps both methods need to be tried.)

While recording, listen intently for any noises in the house or outside which might be mistaken for supernormal voices on playback. If any such are heard, at once notice the setting on the counter so that on playback you will be able to identify the sound.

Voices tend to come immediately after your own announcement or soon afterwards, and therefore I only wait about ten revolutions of the tape. I then reduce the gain to the former low level and speak again. I rarely make more than six announcements inviting voices and switch off. The playback takes so long and is so demanding that it is unwise to make very long recording sessions.

For technicians: with the microphone set at a distance using shielded cable, a good low-noise preamplifier close to the recorder could amplify signals without increasing hum pick-up.)

2. Diode: Recordings using the diode are made in the same way as those using the microphone. You obviously need a small switchboard enabling a quick change from microphone to diode to be made. The preliminary announcement is of course made in the same way via the microphone with the gain low as before. Immediately you have finished speaking, switch off the microphone and switch on the diode at the same time as you turn the gain up full. Wait about ten revolutions and repeat the exercise. Carry this out about six times and switch off.

(Again, the clever technician might develop methods which do not require resetting the gain control, but perhaps it is best to do just as directed until one has experienced the first voices and knows whether results can be had both ways. Possibly Mr. Sheargold already uses more sophisticated methods, but wants to keep these instructions simple enough for the most ordinary experimenter.)

3. Announcements: You will, of course, develop your own technique. Personally I follow Raudive and commence somewhat as follows: "Good evening, my friends in the Beyond, the date is . . . and the time is . . . and I wonder if there is anyone there (he might have said "here" as on several occasions the voice personalities have indicated that they are right in the room with the experimenter. Read the book **Breakthrough**.) who knows me and who can speak to me using the microphone (or diode) NOW!" (Recording at full gain then continues for about 30 seconds.)

My next announcement may be to call by name for some particular friend who maybe has spoken before. In my experience it is unwise to make long-winded announcements. It seems best for them to be short, clear, concise. Also, it is well to remember that since it seems we are dealing with real entities, we should always observe politeness, and I personally always endeavor to acknowledge with thanks any intelligible voice received. Whilst keeping the sessions bright and cheerful, I would avoid frivolity of any kind as it might be misunderstood.

(Note: In reading the book **Breakthrough**, I ran across one passage in which Dr. Raudive suggests that in calling upon the personalities to speak forth, it may help for the person who calls to be in a particular state of consciousness while he does so. Perhaps it is not sound, but thought that attracts the entities, and it has been shown by José Silva that the best state of mind for telepathy is one of very relaxed expectation. Perhaps one reason why it may take some weeks of experimentation before some people record any voices is that the beginning experimenter is too excited and tense rather than relaxed. A few minutes spent in quiet meditation before starting out might be of great advantage.

4. Playback: Unless you have "beginner's luck" it is almost certain that your first playback will reveal nothing. Play the recording back over and over again until you are dead sick of the sound of your own voice. These supernormal voices are very quick or sharp and rhythmic in character—quite unlike anything you ever heard before. When using the microphone method a voice will sometimes come so very quickly after your own announcement that it is almost like an additional word! Be very careful not to miss those voices which say only one word—probably a name. You will hear male and female voices and whispers. Those whispers tend to be slower in utterance and it is usually possible to determine the sex of the speaker. It is impossible to generalize, but if, for example, you carry out all the outlined procedure every day for a week, yet cannot hear anything at all, I would say that almost without a doubt you are just missing voices. It is useless to go on messing up huge totals of recordings yet hear nothing, and I would suggest it is better to spend a week listening to the recordings made during the first week than to continue recording if nothing is heard. I understand that Dr. Raudive himself considers that five minutes, recording involves one hour's listening! Yes, the job is time-consuming!

5. The tape recorder: any reasonably decent recorder will do provided that there is reasonable amplification available, and that the recordings are free from excessive mains hum when the recording is at maximum gain. People have, to my knowledge, successfully recorded voices on cheap cassettes, but the better machines would have the advantage on playback. I have also discovered that some of the older vacuum tube recorders made ten years ago or thereabouts do suffer from hum to such an extent as to render them useless. At present I am using a transistorized Furguson (presumably an English make). One further word of advice: automatic level control (found on most small cassette machines) is quite unsuitable for the voice phenomena. (If the machine has a switch to turn off the automatic gain control, it might be suitable with that feature off.)

6. Logging the voices: You will find it necessary to keep a lot of the recordings you make. Essential details are: date, time, recording system (i.e. microphone, diode, etc.), rev. counter reading, track no., announcements made, and response, if any. Other details can of course be added if desired. Obviously you must work out your own system for abbreviating these particulars, but without keeping any log it is very difficult to refer back to any interesting response, or what is even more important, to follow any trends that may be observable.

Final remarks: Although some may disagree with me in this, my own experiments have revealed no indication that the voice phenomena has anything to do with spiritualistic mediumship. Dr. Raudive denies that there is any connection and I completely agree with him. This is a straightforward electronic experiment.

P.S. By Mr. Sheargold, October 12, 1972:

"I would like to make it clear that I personally am not committed to any one explanation of the voices. They may be the voices of the deceased, or they may be due to some as yet unexplained action of the subconscious mind. That they are a fact in nature is beyond doubt. Our first aim should be to improve their strength and clarity by the development of more suitable equipment.—R. K. Sheargold.

THE DIODE CIRCUIT: Note that the circuit must be fully shielded (enclosed in a metal box, except for the four inch stiff wire antenna.) to avoid hum from the power lines and excessive interference from radio stations. Any (external) wiring employed must similarly be shielded. (The shield or box is connected to the ground or to the chassis of the tape recorder as is usually the case with the shield of the microphone cable—within the machine.

Components:

A. Antenna—four inches of stiff wire insulated from box B and extending outside it.

B. Metal box large enough to hold other parts easily. Obtained in electronics parts store.

C. Radio frequency choke—about 2.5 millhenries. Cost $.69

D. Germanium crystal diode, such as type 1N191 from electronics store. Cost $.35.

E. 100,000 ohm resistor. Cost $.15; may be 1/4 or 1/2 watt size

F. Center lead of shielded cable. Connect to microphone jack or high gain input of tape recorder. Cable should be no more than 12 inches long to avoid hum.

G. Shield of cable. Connects box B and grounded side of tape recorder input as well as resistor and choke as shown.

Conclusion—Significance of the Voice Phenomenon: This article, and the books and record that have been mentioned may lead many people to repeat these experiments. Methods will be improved. More unmistak able voices may come through with solid evidence to show that conscious life goes on long after the physical body is gone. If so, the "dead" do live on and we may all be together again in the future, and perhaps, forever. We may meet the people whom we have helped, and those we have hurt, and if by act or omission we have been responsible for the suffering of others, we may have to bear that responsibility, and even experience the suffering we have caused. The same may also hold true for the joy we have created. These experiments may be part of the breakthrough of the century; part of the dawning of Aquarius.

THE FDA ATTEMPT TO STOP VITAMINS: Recently the FDA, which is supposed to protect people against bad food and drugs, has been making raids on suppliers of vitamins simply on the grounds that the vitamins are called "food supplements," but are contained in tablets which are larger than the "minimum daily requirements." The full details of obviously outrageous attacks (without even telling the manufacturer of the new "requirement") are explained in the September issue of **Prevention** magazine. (Obtainable for $.60 from Rodale Press, Emmaus, Pennsylvania 18049). I suggest you read it and/or write to FDA commissioner Charles C. Edwards (or call him person to person) or contact Hon. Paul Rogers, chairman of the subcommittee on public health and ask him to act positively on bill H.R. 2323, which would prevent such abuses. Sorry I do not have more space to go into this, but it is important.

There is just a chance that McGovern may not be elected next month, but if that happens I feel that Nixon will not be long in office. There is talk of impeachment already. Jean Dixon (who is sometimes accurate) says she feels he will win but he will not finish his term and he will not die in office.

WTAO-FM in Murphysboro, Illinois (618 687 2000) is owned by some people who write to me that they are "very interested in the occult, psychic healing, mind control, yoga, etc." Bill is your man there. Contact him if you have good material. WTAO wants to broadcast words to expand consciousness!

ALTERNATIVE ENCOUNTER FOR INTENTIONAL COMMUNITIES

Charles Cook and Don Leveredge were professional encounter group leaders for schools, corporations, and other establishment organizations, but they found that their successes were lost in an ocean of counter-pressures as participants returned to the establishment world. Now they have decided to travel and give their services to non-commercial work in groups and communities with people seeking finer life styles in New Age ways. You can contact them at 429 Norton Parkway, New Haven, Connecticut 06511 (203) 562-8817) They are willing to arrange fees on basis of ability to pay; even free when required.

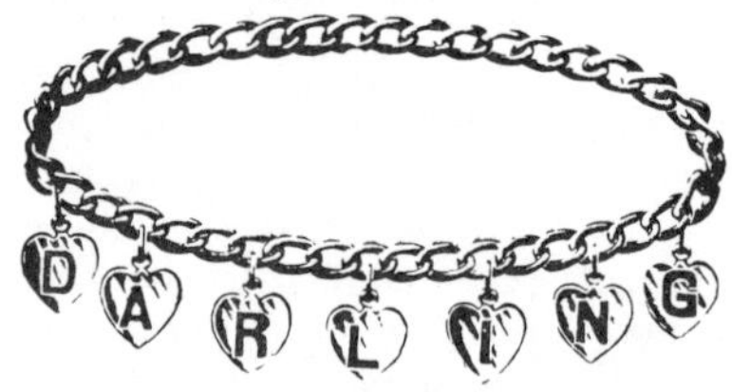

CHAIN LETTERS: Being aware of the power of psychic things, some of you may have wondered what to do with chain letters which promise money or good luck if you send them out to 20 new people, and suggest terrible things if you don't. I receive about four a year, and I do not send them on (although I need the money).

I recently received the following letter:

THINK A PRAYER

"Trust in the Lord, acknowledge him with all your heart, and he will light your way.'

"This prayer has been sent to you for good luck. The original copy came from the Netherlands. It has been around the world nine times. The luck has been sent to you. You are to receive luck within four days after reading this letter. It is no joke. You will receive it in the mail. Send twenty copies of this letter to friends, who think need good luck (sic). Please do not send money. Do not keep this letter. It must leave within ninety-six hours after your receipt of it. A U.S. Officer received $1,000. Don Elliot received $60,000, but lost it because he broke the chain. While in the Philippines, Gen. Walsh lost his life six days after receiving this copy. He failed to circulate the copy; however, before his death, he received $775,000 he had won.

"Please send 20 copies and after that see what happens to you on the fourth day. Add your name to the bottom of the list when copying this letter and leave off the first name."

There followed a list of 21 names, not known to me, and as there seems to be no reason to repeat them as given in the text, I will not do so. I read the letter Saturday evening. To comply with the instructions, I would have to send it out tomorrow, which is the day

I hope to start mailing this newsletter. Good luck has always come to me in the mail—that's what has kept ARF going thus far—so there is no way that the chain letter could be blamed for the luck. The sad thing is that veiled threats are used to get people to continue it, and that tends to be true and effective, but only when a person dwells on and envisions the threatened things happening to him, just as good luck tends to come to a person when he continually expects and envisions good things happening. Anyone who has taken the Silva Mind Control course will understand the factual basis of that claim. To others, it may appear superstitious, but I don't think it is. So, whether or not this newsletter goes out tomorrow, and whether or not I have "broken the chain," I am expecting good luck and I'll get it.

A PARAPSYCHOLOGICAL NOVEL

2150 is the name of a new book by Don and Thea Plym of the Macro Development Center. It is subtitled, **"The Macro Love Story,"** but is a kind of science fiction account of the future as seen through the eyes of a young student in 1970 who experiences the world of 2150 by a sort of time translation that occurs each night as he goes to sleep. The experience is a deep one, going way beyond the science-fiction outlook to combine recent research with the social concepts of a whole New Age. The book is being distributed by the Macro Development Center, and I would recommend it as fascinating and important reading for each of you. The book, 280 pages, sells for $1.50 (add $.25 postage, etc.) and can be had from Macro Development Center Box 26582, Tempe, Arizona 85282.

COOPERATION & THE FUTURE

One of the solid foundations of the future consciousness will be co-operation. Perhaps the best training ground for co-operation is food co-op's. This movement away from supermarkets and conglomerate control towards local neighborhood control of food supply and distribution has about 780 centers nationally. A listing of these in all states is available from FOOD CO-OP Directory c/o AFSC Room 370, 407 Dearborn Street, Chicago, Illinois 60605 for $.50. If you do not have any in your area, you can write for a pamphlet, FOOD CO-OP's, HOW TO START THEM AND MAKE THEM PROSPER put out by Food Conspiracies, 165 W. Harvey Street, Philadelphia, Pennsylvania 19144 for $1 or available at Leaves of Grass Bookstore in Philadelphia. For food co-op's in Philadelphia, call the Ecology Food Co-op, 36th and Race Streets, at (215) BA2-5142.

Food co-op's work at creating an alternative to one aspect of economy. But many other groups are interested in working out a more comprehensive change of life-style. There is far too much already in print about urban and rural communes and villages to add more. We can point out some of the best and most up to date resources though:

Communities Magazine, a journal of co-operative living from Twin Oaks Community, Rt. 4, Louisa, Virginia 23093, costs $1.

It is printed every 2 months and has a 14 page section dedicated to putting people looking for communes in touch with the appropriate group.

Mother Earth News and Lifestyles from 1899 Hubbard Road, N. Madison, Ohio 44057. Each cost $1 and have sections devoted to contacts for communes and individuals.

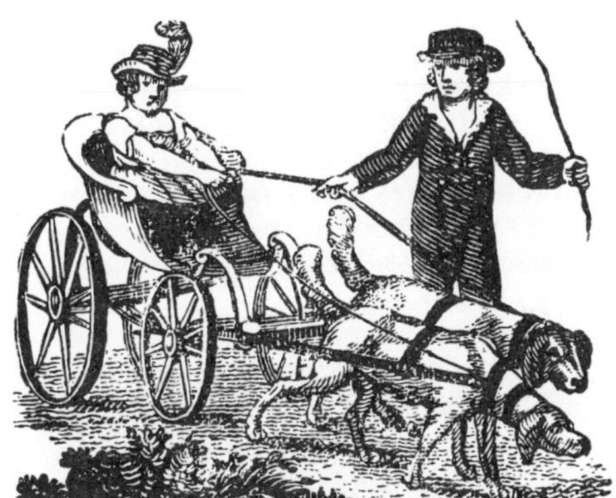

ECOLOGY TRAVEL CLUB

Some people believe it is dangerous to pick up hitch-hikers because it is not easy to be sure whom you are picking up. Yet, hitch-hiking is very sound ecologically, as it causes a minimum of pollution.

A possible solution to the puzzle would be to organize an Ecology Travel Club, which would have a central office to which people could send a photograph, name and address and a small membership fee. In return for that, they would get a large plastic ecology emblem (earth, sun, and sky in three colors—reflecting material for easy night visibility). Imbedded in the plastic of the sign would be the picture and identification of the traveler, so that the driver would know whom he is picking up. In addition, there could be gas stations or restaurants along the road where E. T.'s could wait for rides, and these could be identified by a similar emblem. The traveler and driver could, if desired, both sign the guest book which such stations might keep. And, of course, the people might have a bite together before they take off in the car so as to get to know one another if they felt like it.

That way, hitch-hikers could come to be known as ecology travelers, and the whole thing would get a better slant.

To get such a nationwide club started, someone or a few people would have to advertise the idea and spend time writing about it and getting it together. Beyond that, it should pay for itself.

TELEVISION SPONSOR'S DIRECTORY: Not watching much TV, I never felt the need to know who it is that is responsible for the quality of the programs which are produced. However, a lot of people want to know so that they can write directly to the sponsors with complaints or suggestions.

A complete letter writer's guide for that purpose is available from the Everglades Publishing Co., Everglades, Florida 33929. The directory is published four times a year and costs $3. It gives the name of the president of the company and his office address for every company that advertises on national TV, and also shows which companies own the others.

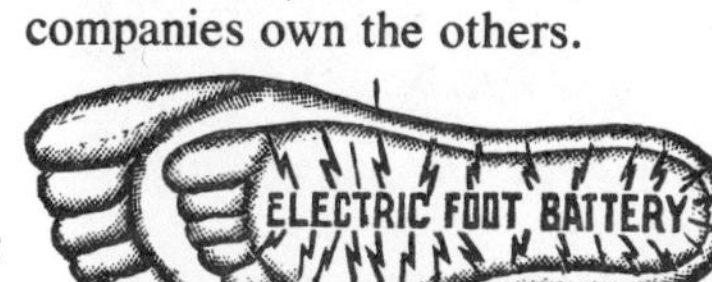

NEW TYPES OF HEALING: On the way to Miami, I stopped and visited Jimmie Scribner in Greenville, South Carolina, after seeing a newspaper article which said he had developed a kind of electronic board which has power to heal people who

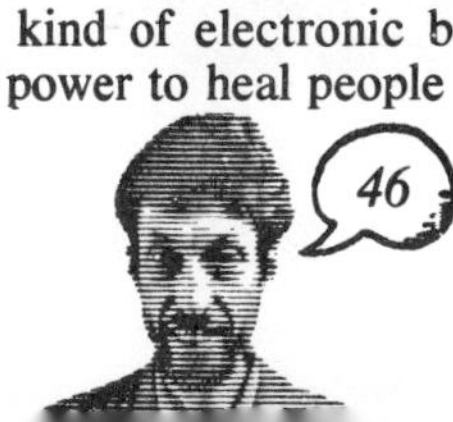

are sick with all kinds of serious illnesses. On arrival, I learned that the device was for sale for $75, and that thousands were being ordered. I heard that doctors were finding them to be very helpful in some almost hopeless cases. The boards were sold with a three month money-back guarantee, and although I did not have the money with me, Jimmie allowed me to take three of them with me for use by my mother, myself, and a friend who has a great deal of interest in this thing. All day, his office was flooded with people coming to buy the new healing device. Because of the large number of people there, it was difficult to talk to Jimmie long enough to get a satisfactory scientific explanation of the theory behind the solarama board, except that it consists of two metallic strips which are connected to opposite sides of the 110 volt A.C. power line, and the strips are attached to a strong asbestos fibre board cut to 2′ x 2′. Between the strips, a special powder is deposited, which is a very finely divided mixture of different metals. As the current flows through the powder, it produces millions of tiny electric arcs. The claim is that these arcs then emit electrons which penetrate the mattress (the board is placed under the mattress of a bed), and in some way return body cells to a healthy state. So far, I have not been aware of any special benefit from the device, but I will give it a longer trial period than is recommended by its inventory (three to six weeks). Of course, not being ill, I don't know what to expect, but was told that I would notice a decided difference in the way I would feel.

My mother is also testing the device, which she is hoping will help with a serious arthritic condition in her hip, and it will be easy to ascertain if that is improved or not.

"CAN GOD FILL TEETH?" is the name of Daniel Fry's book about Reverend Willard Fuller's unusual healing ministry. According to the book, and to the information that the Fullers gave me, for the last ten years, Brother Fuller has had the very unusual gift of being able to pray for the healing of people's teeth and obtain fantastic results in a very high percentage of cases. A figure of 80% is often mentioned.

I was able to visit Reverend and Mrs. Fuller in Jacksonville, Florida, and they gave me some books to read about their work (and I was able to tape an interview with them), and after reading the books, I was able to see them again and have Brother Fuller pray for the healing of my own teeth. As he ended the prayer (during which I had been told to relax and close my eyes), he placed his hands on my jaw, and that moment, it seemed as if a bright light were suddenly made to shine on my face. As he removed his hands, the light and warmth were gone. I opened my eyes, and there was no evidence of anything that could have caused the light in an external way. But the main question is whether or not my teeth would undergo a change as a result of Brother Fuller's work. I am told (and have read) that some of the healing is immediate—at other times it comes weeks later, and first appears as a white spot on a tooth. I should buy a small mirror and watch for the healing to begin. Expectation is probably important. In some cases, according to witnesses, it has been possible to watch the healing actually take place.

Fry's book, **Can God Heal Teeth** is published by CSA Press, Lakemont, Georgia 30552. The book gives names and addresses of people who witnessed and experienced healings of teeth. Typical of Reverend Fuller's work, there is no price on the book. The Reverend does not charge for his services, but is supported by whatever people happen to send him. He also prays for people who send in their request by mail, and I am told that many of them are healed. He also heals other things beside teeth.

Reverend Willard Fuller's address is: 4375 Confederate Point Road, Jacksonville, Florida.

When I experience a change in the condition of my own teeth, I will report to you about it.

ASTRAL TRAVEL EXPERIMENTS are being conducted by the American Society for Psychical Research (ASPR).

Here at ARF, we are still hoping to have an experimental class in astral travel, and a few people are reading the two books which are a preliminary requirement. They are **Journeys out of the Body** by Monroe, and **The Projection of the Astral Body** by Muldoon and Carrington, published by Weiser, 734 Broadway, New York City.

In the later mentioned book, there is a clear reference to the fact that the astral body is believed to weigh about two ounces, and I have suggested to ASPR that they construct the type of scale which can monitor the weight of the person who is supposed to be projecting, so that it may be possible to distinguish in that way between mental projection (or ESP) and the actual projection of an "astral body." The answer from the ASPR was a refusal to try weighing the body—too busy!

KIRLIAN OR "HIGH VOLTAGE PHOTOS": Readers are by now aware of the idea of photographing a kind of aura which surrounds all objects and which has some relationship to the auras that psychics seem able to see. Whether or not the high voltage photographs are really "aura" pictures or some other effect, everyone seems to agree that they vary with emotional changes as well as health conditions, and that this is a very interesting field of research.

Up to now the low cost way to make these photographs has always involved producing the high voltage with devices that have vibrating contacts which wear and cause unexpected changes.

Much better methods are simply described in a book, **High Voltage Photography** by H. S. Dafkin, 3456 Jackson Street, San Francisco, California 94118 ($6).

SCHOOL OF LIVING: The School of Living is one of the oldest rural communal groups that welcomes people to come and learn to start their own rural communities. I have been there and it's great. They publish a newspaper and have asked us to tell you about it as follows:

"Increased awareness and higher consciousness are essential to the New Age.

"So is action for community and social change. All are related to seventeen Universal problems of living that are defined and discussed in terms of action solutions in bi-monthly Green Revolution. Join its wholistic approach. Send $6 for a year's subscription or send a stamped self-addressed envelope for a catalogue to: School of Living, Box 129, Route 1, Freeland, Maryland 21053."

1. Negative – Non Malignant

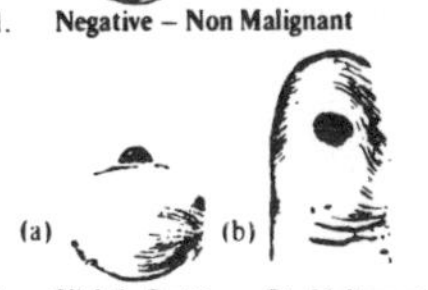

2. Slightly Positive – Pre-Malignant

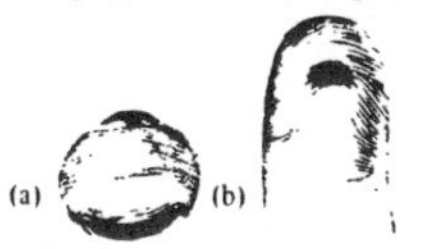

3. Positive – Undetected Malignancy

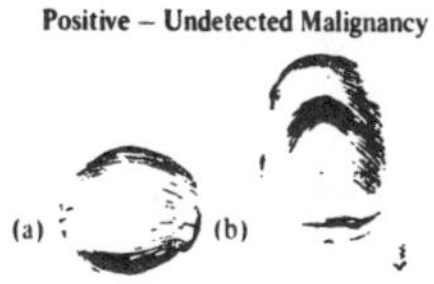

4. Very Positive – Clinically detectable Malignancy

A simple, inexpensive test designed for the mass screening of the general population for the early detection of malignant and pre-malignant conditions. An accuracy exceeding 95% has been indicated for the slow growing types of malignancies. This test often gives positive indications 2 to 5 years before the cancer becomes clinically evident. It is less accurate for the fast growing types of malignancies — an accuracy of 60% has been indicated. Laboratory tests may be used for a more accurate diagnosis.

Cancer is a result of an inability of the body to properly metabolize protein. In this screening test we are observing the effects of variations in protein content of the blood. When the blood contains an adequate amount of protein, it has a consistency with enough "body" to form a well-developed hemisphere. When the blood contains an inadequate amount of protein, its consistency lacks "body" and the blood spreads over the finger.

False positive indications are possible. A person can have low blood protein without having cancer. Some other conditions can cause the blood to be excessively thin. It is nearly impossible, however, for a person to have cancer without having deficient protein metabolism. In most cases with a positive indication, immediate concern and treatment for protein deficiency is indicated. Additional protein in the diet is contra-indicated since the body has failed to properly utilize the protein that has been available. The most frequent causes of inadequate protein in the blood are a lack of pancreatic enzymes and a mucus blockage of absorption in the small intestine.

False negatives are also possible if the body is dehydrated. In terminal cancer, the blood is often so changed that fluids build up in body cavities or are lost from the body. As the blood loses fluid, it becomes thicker and false negatives can result.

TRUE EARLY SIGNS OF CANCER

The true early signs of cancer are the signs of inadequate protein metabolism and may be called pre-cancerous symptoms. These are:

1. **Gas in the stomach or bowel.**
2. **Sudden weakness of the eyes.**
3. **Tired feeling most of the time.**
4. **Muscle weakness and cramps — first in the back, then in the chest.**
5. **Extreme mental depression.**
6. **Sudden change in hair color or texture.**
7. **Development of various hernias (only in slow growing tumors).**

A PERSONAL NOTE: Recent times, my own health has been excellent (better and better) but the dentists at the University of Pennsylvania tell me that my gums are receding. They say they are not able to cure this problem, and that merely to check it will require treatment needing 20 minutes a day, possibly for the rest of my life.

The Silva Mind Control people and some spiritual healers tell me that it can sometimes be cured by prayer, meditation, and the mind control method. I would like to overcome the problem in that way, and would therefore like to ask that any of you who are acquainted with distant healing, or who know answers to the difficulty, should keep me in mind in that respect. Your help will be appreciated, and I will have the dental school check for signs of progress.

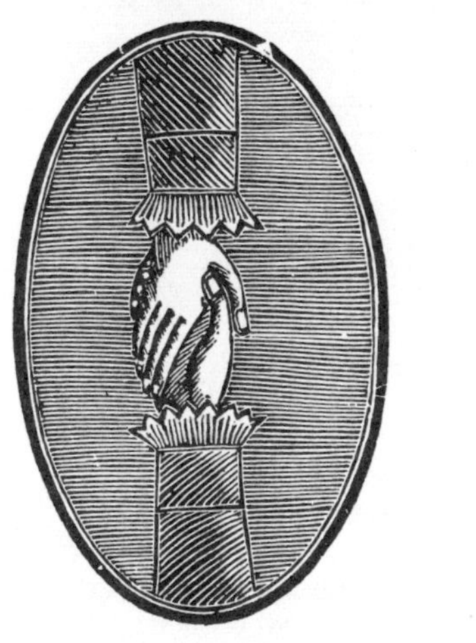

COMIC BOOKS FOR THE NEW AGE! During the three years' life of ARF, a tremendous amount of new information has come in, and much of it has been verified or proven false in some way. Trouble is, it is not always gotten out to the public very well. Some of the best verified material (such as Laetrile for cancer) is not published by most publications because they simply cannot believe the enormity of the swindle, or the information is just too "far out" for most people to look at in a serious form.

So, enter the comic book! Underground comics seem to be a medium which could be used to tell the truth in a style which people would be ready to read. The story can be fictionalized enough to make a connected and interesting narrative. At the end, however, it is pointed out to the reader that what they have read was actual fact—stranger than fiction—truth. Names, addresses, and telephone numbers could be given to enable the reader to get scientific corroboration, if that is desired. Those wishing to seek further could do so.

Already, a couple of comic book artists have answered an ad of mine and I am encouraging each one to write on the field that is of most interest to him.

A BRAIN WAVE SLEEP/DREAM EXPERIMENT: Just after returning from Miami, I developed a sore throat, and not having time to be sick, I decided to try a healing method mentioned by some people at a mind control graduate group. The method consists of placing an oscillator near the source of pain, so that a ten cycle per second or Alpha wave frequency stimulates the body area to accentuate healing. I connected my J & J Enterprises brain wave signal generator to the brain wave machine and set it at 10.5 cycles per second. Then I connected the output of the brain wave machine to an amplifier, so that the 10 cycle sound (actually a modulated white noise, as the ear cannot hear 10 cycles per second) was clearly audible, and I placed that near my head, and went to sleep.

I woke up just a few hours later, feeling that I had been dreaming continuously, but had not had a restful sleep at all. I could remember parts of all sorts of dreams. Later, I thought about what had happened, and realized that hearing the Alpha sound may have kept me from going into the normal Delta brain wave state, which is the first part of sleep.

I realized that an interesting dream experiment could be devised by making a tape recording which would start with about 90 minutes of Delta waves to induce deep sleep, and then would rise up into the Alpha range where dreaming occurs. Perhaps the brain would follow those sounds and dreaming would begin when the recording went into Alpha. After about 20 minutes of Alpha, the tape could then awaken the sleeper, telling him to record his dream, for which the tape recorder would be ready.

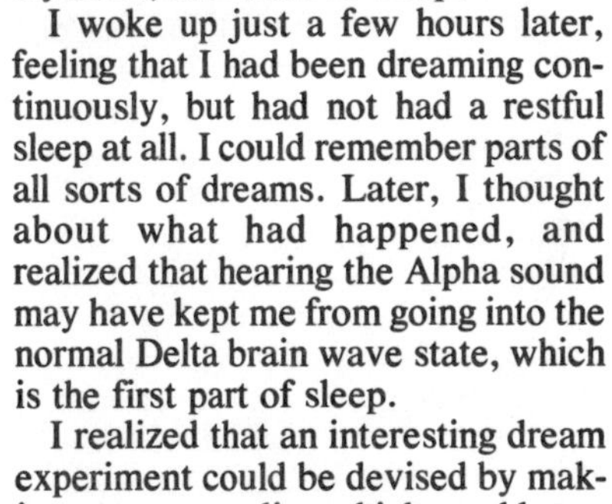

THE SECRET LIFE OF PLANTS!! This is a new book by Peter Tompkins and Chris Bird. (Harper & Row, $8.95) I am ready to predict that this book may become a best seller, and it will at least be as much of a mind opener as **Psychic Discoveries Behind the Iron Curtain.** It is not written in quite the journalistic style of the other, but it is packed with far more details and will help you to understand the mess we are in today. Here is a summary: written mostly by

Bruce Dickson:

This 400-page book begins with a comprehensive report on all the current research on new dimensions in plants now opening to scientists around the world. The first pages entitled "Pioneers of Plant Mysteries," with everyone from Goethe to Sir Jagadis Chundar Bose of India (who out-Backerstered Cleve Backster years ago but got little publicity).

By this time, the lay reader's mind is so open to new possibilities and cosmic implications of so many separate experiments, that it would have been a crime to stop the book here. So armed with the same enthusiasm with which the first sections of the book are written, the authors plunge into the sea of New Age study, and try to show how we could begin to utilize some of this vast untapped wealth. The rest of the book is, if possible, even more absorbing than the first part about our vegetable buddies.

Some of the things covered are: Interstellar bio-communication, Lakhovsky's multiwave oscillator, how artificial lighting creates hyperactivity in children, of John Ott's Environmental Health and Light Research Institute, Kirlian photography, etheric and other bodies, auric fields, organic farming, food additivies, effects of poisons in agriculture, transmutation of elements, Wilhelm Reich, dowsing, the Hieronymus machine, radionics, telekinesis, and Findhorn Gardens.

Most of these appear in a book from a major publisher for the first time, and all of it is written with a consumate, loving craft to make it as readable as possible. And it's beautifully indexed for all our hardworking researchers. Long bibliography, too.

On page 253, the authors quote Dr. Roger J. Williams, discoverer of pantothenic acid and director of the Clayton Foundation - Biochemical Institute in Texas; the first biochemist to be elected president of the American Chemical Society. He gives these figures in relation to the birth of:

RETARDED CHILDREN: In 1952, 20,000 retarded children were born in the U.S. By 1958, it was 60,000 and by 1964 it was up to 126,000. By 1968, over 500,000 children per year were born retarded.

I have tried to get verification of these figures from the Pennsylvania Association of Retarded Children, but it has not come through.

THE POSITIVE EFFECT OF PRAYER ON PLANTS
By Robert N. Miller, Ph.D.

In his book **The Power of Prayer on Plants**, Reverend Franklin Loehr describes many experiments in which the growth rate of plants was apparently accelerated as much as 20 percent when individuals, or groups of people, "prayed" for plants in a systematic manner. The recommended manner of "praying" was to visualize the plant as thriving under ideal conditions. The growth rate of the plants was determined by measuring their height with a ruler at relatively long intervals ranging from several days to several weeks.

Although the results reported seemed evidential, an element of doubt remained because the tests were not conducted by scientifically trained personnel and because the method of measuring growth rate was relatively crude. In 1966 Dr. H.H. Kleuter of the United States Department of Agriculture developed a very accurate method of measuring plant growth by using a rotary transducer connected to a strip chart recorder. Through the use of this device, he was able to measure the growth rate of grains and plants to an accuracy of a thousandth of an inch per hour.

In January of 1967 I devised an experiment to attempt to measure the effect of thought upon remotely located plant life using this highly precise method of measuring growth rate. My volunteer subjects were Ambrose and Olga Worrall. Ambrose, formerly an aircraft engineer with the Martin Company of Baltimore, Maryland, has been credited with the healing of hundreds of individuals. His wife, Olga, also has the reputation of being an effective healer. Many of their case histories are documented by the reports of both physicians and patients in **The Gift of Healing** and **Explore Your Psychic World**.

The selection of a plant suitable for the experiment turned out to be more of a problem than was anticipated. Initial experiments were performed on bean plants because they are sturdy and grow rapidly. A diagram of the test arrangement is shown in Figure 1. The tip of the plant was attached to a counterbalanced lever arm which, in turn, was connected to a Model 33-03 Brush Metripak Angular Position Transducer. The slightest motion of the lever arm was converted to an electrical signal which caused a deflection of the indicating pen on a strip chart recorder. A wide range of accuracy could be achieved by adjusting the amplification of the recorder and the chart speed.

The bean plants, when watered every 24 hours, had a rapid growth rate. However, it was found that the strip chart recorder did not measure the growth rate because the new growth always occurred at the tip of the plant above the point where the lever arm was attached. After experimenting with a number of other plants, it was found that ordinary rye grass has all the necessary characteristics. It is sturdy, and grows rapidly, and the new growth occurs at the bottom of the blades. Consequently, a lever arm attached to the top of a blade of rye grass will measure total increase in length.

Each test was started by placing 10 rye grass seeds in a 25 millimeter plastic beaker filled with a fertile soil. The seeds were planted 1/4 inch below the surface of the soil. The seeds were watered every morning with five millimeters of water. They began to sprout after two days and, after four days, the blades of grass were long enough to permit attachment of the lever arm. A small piece of Scotch tape provided the connecting link. The strip chart recorder was adjusted to a chart speed of one inch per hour and the amplifier was set so a pen trace with a 45 degree slope represented a growth rate of 0.100 inch per hour.

A number of preliminary experiments were conducted with different sprouts of rye grass. The grass did not grow as rapidly in the dark as it did when exposed to fluorescent lighting. The growth rates varied between 0.002

and 0.10 inch per hour. Under the constant conditions of lighting, temperature, and watering frequency selected for the test, the growth rate was approximately 0.006 inch per hour. At no time did the growth exceed 0.010 inch per hour in any of the preliminary experiments.

When Ambrose and Olga Worrall visited Atlanta in the fall of 1966, I showed them the plant experiment and they agreed to cooperate in a distant prayer experiment when they returned to their home in Baltimore, Maryland, some 600 miles away.

Before the experiment began in January of 1967, the growth rate of a new blade of rye grass had been stabilized at 0.00625 inch per hour. During the night of January third the trace on the strip chart recorder was a sloping straight line indicating a constant growth rate. The straight line continued with little or no deviation during the next day. At 8:00 P.M. on the evening of January 4th, I telephoned from Atlanta to Baltimore to ask the Worralls to hold the seedling in their thoughts at their usual 9:00 P.M. prayer time. One hour later they "prayed" for the plant by visualizing it as growing vigorously in a white light.

The next morning I carefully examined the strip chart recorder trace, All through the utterly convincing evening and until 9:00 P.M. the trace was a straight line with a slope which represented a growth rate of 0.00625 inch per hour. At exactly 9:00 P.M. the trace began to deviate upward and by 8:00 A.M. the next morning the grass was growing 0.0525 inch per hour, a growth rate increase of 840 percent. (Instead of growing the expected 1/16 of an inch in a ten hour period the grass had sprouted more than 1/2 inch.) The recorder trace was continued for another 48 hours. During that time the growth rate decreased but did not fall back to the original rate.

During the experiment the door of the room had been locked, the temperature was constant at 70 to 72 degrees F, the fluorescent lights were on continuously, and there was no known physical variable which could have caused any large variation in the growth of the rye grass.

The dramatic results of this experiment—an eight-fold increase in the growth rate of the rye grass occurring while distantly located "agents" were trying to increase the growth rate—suggest that this sensitive experimental technique can be effectively used to accurately measure the effect of mind over matter.

Dr. Robert N. Miller is an industrial research scientist who also directs research for the Atlanta, Georgia, chapter of the Spiritual Frontiers Fellowship. He received his B.S., M.S., and Ph.D. degrees in chemical engineering at Ohio State University.

He was formerly a professor of chemical engineering at Georgia Tech, where he taught courses and directed research in the fields of metallurgy, high polymers, corrosion, fluid flow, and heat transfer. He is a member of the American Chemical Society, National Association of Corrosion Engineering, the Electrochemical Society, and is listed in **American Men of Science**. He has authored more than 20 scientific papers and is the holder of two patents.

PRAYER HEALING THROUGH BIOFEEDBACK It is obviously time for us to stop asking if prayer is real and to start learning how to use it in our lives. How can we learn to be effective in prayer?

Imagine for a moment that the above experiment (article is from **Psychic** magazine, Box 26289, Custom House P.O., San Francisco, California 94126, $5.50/year) were to be modified slightly so that instead of taking hours to measure the growth of the plant, we had a meter in front of us which could instantly indicate how fast it is growing at every moment.

As the Worralls pray, we see an immediate rise of the meter needle; it rises from 1 (normal growth rate) to 8.4, where it remains all night.

Now, we connect that meter to a different plant, and ask YOU to pray for it. You are permitted to watch the needle while you pray. What happens? Perhaps nothing. You alter your mental vision, trying different approaches—then, suddenly, the needle moves a bit and stops. What did you do? It takes some time before you can make the needle rise again, but eventually it does—you remember the kind of thoughts that make it go higher.

Each day, you spend an hour or two developing the kind of thought-attitude that makes the plant grow the fastest until after a number of weeks, you are able to do far better than you did to begin with. You have been using the biofeedback method to develop your own prayer healing ability!

AN INDIAN PRAYER: In seeking advice on ways to bring an end to the war, I was asked to publish an Indian prayer for racial reconciliation from the **Hermetic Philosophy**, and so I agreed:

"God of all nations, we pray Thee for all peoples of Thy earth; for those who are consumed in mutual hatred and bitterness; for those who make bloody war upon their neighbors; for those who tyrannically oppress; for those who groan under cruelty and subjection.

"We pray Thee for those who bear rule and responsibility; for child races and dying races; for outcast tribes; the backward and the downtrodden; for the ignorant, the wretched and the enslaved.

"We beseech Thee to teach mankind to live together in peace, no man exploiting the weak, no man hating the strong, each race working out its own destiny unfettered, self-respecting, fearless.

"Teach us to be worthy of freedom, free from social wrong, free from individual oppression and contempt, pure of heart and hand, despising none, defrauding none, giving to all men in all their dealings of life the honor we owe to those who are Thy children, whatever their race, their color or caste."

2,000 GET SILVA MIND CONTROL FREE: The Silva Mind Control Method is to be taught to all students free of charge. That is José Silva's deep desire. It has begun to happen here in Philadelphia!

All 2,000 students of the Hallihan Catholic High School for girls were given the full 48 hour Silva Mind control course without charge as a special experiment. José himself came to help with the planning, and ten instructors came from various places to teach the course. All other school programs were abandoned for the week and seven hours a day were given over to learning the mind control method. I understand that all the classes went very well, and students were generally enabled to do the psychic readings at the end of the course. Now, some kind of psychological testing is being done to see what the results might be, but I have no information on just what that testing consists of. In time, I expect there will be a report. All the school's teachers took the course as well, and the principal had taken the course previously, which is what started the ball rolling. News of these events appeared in the front pages of both the city's major newspapers.

NATURAL BIRTH CONTROL
An article on birth control, published in **Neue Revue**, written by Erich Wiedemann, is entitled "Plant Sap instead of Abortion" (translation). It tells of the journey of a Western doctor to the jungles of Brazil, where he finds a native tribe which uses a mixture made from plants as an "after the act" birth control preparation. The article does not reveal the nature of the preparation, but leaves the reader assuming the doctor, who is now back in Germany, will soon find that out. Perhaps he will. Perhaps it is already known. I am referring to Laetrile, which I have been told will tend to prevent pregnancy when taken in sufficient quantity, because it attacks the trophoblast cell, which produces the opening in the wall of a woman's uterus where the fertilized egg can take hold. I would like to telephone Dr. Krebs (at the John Beard Memorial Foundation in San Francisco) and ask him about it, but I am afraid I will have to wait until my finances are in better shape. I have heard, however, that some women in Hawaii's island of Maui use papaya seeds (rich in Laetrile) in that way. I sent that information on to Dr. Krebs, but have not received a reply, and have the feeling that he would prefer to wait on that question until the government stops interfering with the use of Laetrile for cancer therapy.

GROWNUPS FORBIDDEN is the title (in translation) of the other German article published in **Stern** and written by Gordian Troeller and Claude Deffarge. I have just read through the English translation, and would recommend it all.

HARPER'S MAGAZINE for November, 1972, will carry an article by Christopher Bird telling about many of the most important advances in scientific research such as Cleve Backster and beyond. Much of the work he has delved into has been neglected for years because people were unwilling to look at it before the Russians and Backster made their discoveries. Read that article.

HEALING METHODS: In recent times, I have been checking out two types of healing. One was the "solarama" board. In that respect, I am sorry to report that I do not yet have any clear evidence of benefit from use of the device.

I also received prayer healing for tooth problems from Reverend Willard Fuller in Jacksonville, Florida, of whom it is said that through his prayer and laying on of hands, many have experienced miraculous cures. I am sorry to say that I have not so far observed any such cures in my own case. I do not, however, have reason to doubt the veracity of other accounts.

GROWNUPS FORBIDDEN: The Murias, 200,000 people, live in the territory of Bastar, south of New Delhi, India. Among them, children of each colony form their own culture, and live together freely in their own house. There is no adult supervision, and no sexual restrictions other than the rules the children make themselves. One of those rules forbids any couple from sleeping together more than three nights in the belief that to do so would lead to pregnancy. In spite of considerable sexual freedom, only four per cent of the girls get pregnant before marriage (which is usually after 18). There is almost no crime found among Murias; they are a peaceful and unusually happy people. I found this information in a recent article in the German magazine **Stern**. The article states that many sociologists, and even the Swedish parliament, are seriously considering whether the experiences of the Murias might not show us a "way out of our threatening sexual anarchy."

PSYCHIC SURGERY IN THE PHILIPPINES
A month ago I was given an article by an Alternative Feature Service writer on the subject of psychic surgery in the Philippine Islands.

The article told of meeting with an American psychologist, who had spent six months in the Philippines learning about this, and who had decided to apprentice himself to the best psychic surgeon he could find and learn the work. It also mentioned various healers who were able to perform surgical operations with their bare hands, using neither anesthetics nor antiseptics on patients who nevertheless felt no pain, and who often recovered from incurable illnesses as a result of the operation.

Best known in the West in Tony Agpaoa. Douglas Voeks studied with him for six weeks and then visited others and finally came to David Oligane in the lowlands of Luzon. David was obviously one of the best, and his surgery was able to cause the body to open up without even touching it. He simply passed his hand through the air six or eight inches above the surface, and the incision was there. Doug was able to observe this and photograph it, both with still and motion pictures.

Even Doug's own mother, who was ill with diabetes, was cured by one of these healers, and the cure was later confirmed by Western medical tests.

To begin with, I shall present a summary of tapes and telephone calls, and then the article which Doug sent me himself.

The first tape is summarized as follows: "I spent 6 months and 11 days in the Philippines. For 6 weeks I was with Tony Agpaoa, and the remaining time with David Oligane and several others. I stayed in David's compound at his mother's house. David spoke to me many hours a day, using the Bible as his only reference book. There were approximately 12 of us in addition to David who comprised the hard core group and worshipped together in the chapel with great regularity. We came together in prayer each day and night at frequent hours, so that we regularly had a 20 minute prayer meeting at 8:00 P.M., midnight, 3:00 A.M., 6:30 A.M., midday, and sunset. Each one would take a turn at leading the prayer meetings, myself included. The prayers were from the heart and genuine, not memorized.

"David knew the Bible extremely well, and when I asked questions he could quote chapter and verse. I asked if he had memorized it and he said, "No, it just comes to me from the Spirit," evidently, as by clairaudience.

"In seeking the answers to healing, I found that the answers do not come as by a learning process, but as a changing of the old nature—or a being 'born again' as it says in the Bible.

"That new attitude is a prerequisite to receiving the power of healing. How do you change the old nature? Love your brother as yourself: do this continuously, consciously, this is the way you serve God—with all your heart.

"The prayers were mostly in native languages, but often they would pray in English so I could learn; no precise form, but always ask for forgiveness of sins, pray for afflicted people, for the world and for its leaders, as well as for people they know personally. Each healer has a banner with symbols, and each is different, as it comes from the Spirit. Sometimes they tie the names of sick people into their banners. The banner is their all-important contact with God and with their spirit protector.

"After being with Tony, I felt that he had suffered under the corrupting influence of the American scene—something for which we are all responsible—we cannot put all the blame on Tony. But I found two healers who were genuine. One was Joe Mercado, who took time with me for mediumship training, and then Brother David, who encouraged me to meditate on answers to my questions.

"One day, I was taken along on a mission to make up the number of seven, which was required. It was clearly felt that the Spirit was leading them to take me along, and I took part fully in the task of the mission which lasted eight days.

"The message of the Espiritistas is always the same. It is the message of The Revelation in the New Testament: The coming of World War III, which is now seen to be almost unavoidable; because the wise men (national leaders) cannot control themselves. They believe that sooner or later the world is in for it. Edgar Cayce's message is considered appropriate, and two books, **The Late Great Planet Earth** by Hal Lindsey, and **Guide to Survival** by Salem Kirban give a similar message here.

"The only way that World War III could be avoided would be for there to be a tremendous and general repentance on the part of all peoples, and that seems totally unlikely at present.

"A Japanese ear, nose, and throat doctor, Ichiro Kudaka, who is a spiritualist, was also there with us 'to learn from the spirit in the Philippines,' and he was warmly accepted by the Espiritistas and was allowed to use his 'Phlaido system' of getting rid of negative powers, which we were also grateful to learn from him. But above all, everyone is amazed at David's power to cause the body to open up without even touching it, and that takes all the doubt out of the 'Doubting Thomases.' There may be some fraud with one operator or another, but David's work is totally honest; when you question yourself later, this type of incision erases all doubt.

"David is also able to use trance states to remove the effects of witchcraft or negative thought from the person who has been victimized by them; sometimes by an actual operation. He says, 'In the final days, the power of Satan increases very much, but the power of God is always a step ahead.'

"This seems true with the rapid rise of witchcraft, or 'psychic attack' that has been experienced in the last ten or fifteen years—also among Americans who come to David, but this is a difficult concept to portray, so if you cannot accept it, just let it pass for the present.

"David can also cause material changes to occur in water and other objects, and when he does so, everyone can see the color of the water change and people then drink that water as medicine and know that it will make them well.

"He can also produce material tablets as medicine (apports) as he did for Stephanie Kramer of Chicago. Some other operators can do this also.

HOW THESE POWERS ARE RECEIVED "Some have received these powers by approaching another medium healer **after** they have fulfilled the requirements of the Spirit in changing their life completely. Then the Spirit protector of the other medium can give these powers to the new healer. Otherwise, powers may be received by what is called 'mediumship planista,' which at first sounds like astral travel, but which actually goes far beyond the astral (or earth planes) to a much higher plane.

"It is as if the person's spirit (and his awareness) leaves his body and travels to Heaven, to other planes, where spiritual gifts are given. Mediums who have received power in that way quite independently all seem to agree in their descriptions of the experience. They describe the higher planes in the same way: large flowers, rooms of operating equipment, etc. (Most of these have never been inside a hospital.) Going to these places seems to depend on having changed your whole life and lifestyle beforehand.

"Already, while on missions, I have received the gift of being able to heal some illnesses by what is described as a spiritual injection, which has healed some things that doctors have not been able to touch. Hopefully, after I return to the Philippines, I will receive something more. Ultimately, I hope to be able to do the operations because a lot of people can use that kind of help.

"It should be understood that these things are not an absolute cure for everyone who goes there. I've seen complete cures, and fractions thereof. Some people who've had a disease for 20 years expect a five-minute operation to cure everything. It's not like that. It is something to be honored and cherished. The person's faith is the most important thing. I've seen back problems and other things cured instantly, but the Spirit is waiting for us to come around also. We're going to have to give before we will receive."

WORLD WAR III

In a speech of David Oligane which Doug sent me, mention is made of the terrible likelihood that World War III was soon to take place, and that each one should change his life in such a way as to be prepared for that. This led us to serious considerations, as it appeared that the healers believed that World War III was certain and unavoidable and close. We then telephoned Doug to ask about this, and (with permission) also taped the call:

Art: What can we do to prevent World War III?

Doug: Nothing we can do. It's a question of self-salvation at this time.

Art: Not a warning?

Doug: Is it unavoidable? Yes, I'm afraid it is because the "wise men" cannot control themselves.

Art: Another topic—what about Harold Sherman's book, **Wonder Healers of the Philippines**; is it true?

Doug: It was true when it was written, but it is not accurate now.

Art: How would you feel if James Reston of the **New York Times** were to come to do a report on this? (He was the reporter who experienced acupuncture in China, and wrote about it in a way that was believed by many here.)

Doug: That would be all right.

Art: Can you give me more information about the danger of World War III?

Doug: I'll send you a tape on it.

WORLD WAR III
(From the tape received on December 13, 1972)

"Any one of the Espiritistas will tell you it's imminent. Read the two books mentioned earlier.

"True negotiations will not be reached. So everyone has to worry about his own salvation. Biblical prophesies are being fulfilled. Espiritistas say their information comes from the Spirit as well as from the Bible. David always preaches the word of God, repentance, Second Coming. The hour is late. I believe the Third World War is unavoidable unless you were able to reach everybody with a message like this and they suddenly became convinced. That's not going to happen. That's why it's not being a fatalist. Even in the Old Testament, people used to run around saying this city is going to be destroyed, and they became very unpopular individuals. Nevertheless, what they said came true, and this is going to come true.

"So now," as David says, "a time to prepare, prepare yourself and your investigation should be based on finding out for yourself how to prepare.

"As far as the Espiritistas are concerned, I feel that the prophecy that they are receiving is concerned almost more directly with their country. This may be a starting place. I think it's going to be a boiling point soon. They have a more worldwide scope involved, which I feel will come later, but it's going on here and there; wars and earthquakes, etc., but I think the Spirit is warning them of what's going to happen in their country.

"A lot of people are predicting it now, and the Espiritistas have been predicting it for a few years. David feels leery about 1974 to 1975. They feel this will be when world involvement will be imminent and, as I observe it, I think that is when their country is going to be involved, and I think the Spirit is getting them ready and they are preaching 99% of it to their own people. They have no contact with outside people except ones like myself who come in to study, and then this is laid on them. They get the warning clairaudiently and see the signs on their banners. It's not a light message, but something they spend full time on. I asked David if World War III was inevitable, and the answer was 'yes.'

"I can only suggest that you make your own investigation. Find other people with spirit protectors (clairvoyants, psychics?); see if they agree or if it seems to be a localized thing for the Philippines. I don't think so. I think eventually it's going to snowball."

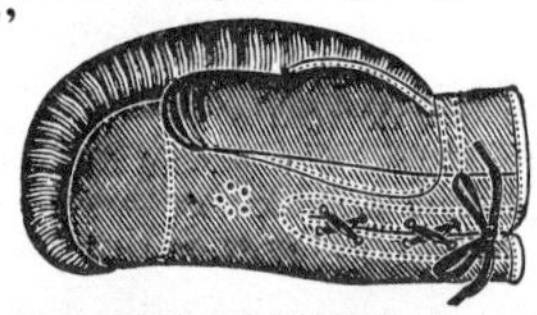

HEALING HANDS by Dutton. This book deals with psychic surgery in England, so I sent a copy of it to Doug. Apparently he read it right away. Dr. Lang is George Chapman's spirit control. Chapman is the medium. It's an actual account of work going on there. Doug refers to it on the tape as follows:

"I believe every word of it. It's so similar to what's happening in the Philippines, except that the spirit protectors are not doctors, or if they are, they are not admitting it. It's more believable to the people if the Spirit is in the name of a saint.

"You're selling God to people who are on the point of changing their lives, and that is the way it's best understood there as it may be better understood the other way in England. I believe Dr. Lang truly is coming through, and there is no way this can be the product of anyone's imagination. I believe that my place of training is in the Philippines, and as long as people are allowed in, I'll continue to go back and learn what I can."

A WORD FROM DAVID OLIGANE (as taped by Doug in the Philippines) ("At this point I would like to quote directly from a talk which David gave. Please excuse the poor English, as David is obviously speaking in a foreign language for our benefit."—Doug)

"It was just the time I was pursuing my studies that I became ill, which is already incurable to medical doctors, until I found myself with the Espiritistas people, which help me to recover my health. This I become to thank God, who gave me that kindness. So I began to search and prove what was going on in this world where I stay. I began to wonder more and more, and my faith became stronger and stronger until I decide to seek the power and grace of God. People coming here every night and day . . . people in every walk of life . . . doctors, nurses, people of high degrees from school come to ask for help . . .

"My friends and my brothers, Philippinos, Japanese, American, German—any kind of people; tomorrow we die and our life is in danger because our life is just like a smoke in the flesh. Time of the Holy Spirit—time to repent. Time now is too short to live on this planet earth. Know my brothers and friends from America, prophesy is already fulfilled. War, nation against nation, earthquake, famine, many sicknesses and a troubled world. No peace in this world is to be achieved because we are now on the brink of coming Third World War. The dangerous weapons made by man to kill man, weapons that could wipe out civilization . . . atomic, hydrogen, oh that germ warfare, the most deadly and dirtiest that can erase mankind. All this weapons is intended to me, to you, and to all mankind. How can we escape this seeking killer? No place to hide. War can explode any time when the wise men cannot control themselves. I mean the government governing the people. This sick world of ours needs help from above. God, the Holy Spirit, the Most High . . ."

ARTICLE SENT BY DOUG VOEKS (reprinted from "Bulletin of the California Society for Psychical Study, Inc.," November 1972. Address P.O. Box 844, Berkeley, California 94701. Price: $2 a year.)

"I've seen some people who, faced with that kind of reality, cannot believe it."

This was how Doug Voeks, a recent graduate of the University of Alaska, who went to the Philippines after seeing a movie of the "psychic surgeons," and who is now in this country for a time before returning there to resume his Espiritistas religious training with Brother David Oligane, introduced this most controversial subject. Mr. Voeks freely admits that, on the part of the medical profession as well as the general public, there exists a complete gamut of reaction, running from varying degrees of acceptance, through neutrality, to utter rejection as obvious fraud. After spending six months in daily close contact and training with the healers themselves, always in search for the "truest men and women" in the field, even those least accessible to foreigners, Mr. Voeks feels qualified to say concerning his teacher: "He proved it to me. IF HE were here he would prove it to you."

Dismissing the theory of mass hypnosis as incompatible both with the actual circumstances of the operation, as well as with the often successful results of the treatment of what are usually considered organic disorders, Mr. Voeks asked the question, "What is it then?" To attempt to answer this question, he said that probably the best thing he could do would be to show the movies to give the general idea. There are lots better movies around—as he was running out of money for equipment, he said—and then to show the color slides, which were better, he felt,since they showed a week by week record of his entire six months' stay. This turned out to be true. The movies suffered from insufficient light, but the slides were really new and different, for those who had already seen a number of movie films, as well as being photographically very good. Answering a question concerning the types of operation being done by Tony Agpaoa and Marcello Heiner, Mr. Voeks commented, "There is no way medically that you could find out much except by standing there with a doctor." He added that none of the healers had received any medical training at all, and that most had left school after the third or fourth grade, and few having finished high school. He emphasized that most of the patients, except those treated by Tony and Marcello, were native Filipinos, who had been given up by regular doctors, or who were too poor to go to them. Commenting on the healers in general, Mr. Voeks said that medical knowledge is not actually needed, as "they're not really in control of their hands at this point, though they can control blood flow and things like that," and control whether or not you will have a scar. No anesthetic is used, there is little or no pain, and no infections develop, he said. "The incision is made with the right hand and the left hand is called "the maintaining force;" when they remove the left hand the body closes."

During his six weeks at Tony's house, Mr. Voeks saw, he believes, between 500 and 1,000 operations, often with good results. After leaving Tony, for reasons he said he would not specify, Mr. Voeks travelled to the interior, in the "lowlands," where those receiving treatment were nearly all native people. There, he feels, he was extremely fortunate to find as a teacher Brother David Oligane, a man of simple life and devout faith, "an all-loving person" who refused to accept any payment for room and board for the four and a half months' stay, saying, "My house is not for sale." Brother David's sect is called "Espiritistas Christiana de Filipinas."

The history of the Espiritistas in the Philippines goes back to about 1850 when a group of missionaries from France introduced it on Luzon. All of the present day healers (numbering about 33 men and 7 women) have come from a small area of this island. The original "medium operator" in the islands is still living at the age of 76, Brother Terte. Mr. Voeks explained that in 1943 the healing force became "material"; that is, visable. Before then it had been entirely "spiritual," or invisible. "The same actions and the same results as with the material or visible healing"; though sometimes, with some conditions, the results were even better with the spiritual type. He later cited the case of a Chicago woman with a diagnosed inoperable brain tumor, who was successfully treated by Brother David with spiritual healing after being rejected by Brother Mercado (at Brother David's chapel) for material healing.

Asked whether the visible part of the healing was mostly for show, Mr. Voeks replied, "Yes, it's absolutely only for show," adding that it depended on the patient; that some patients required it. Needless to say, this placed the entire picture in quite a different perspective from what many of us had imagined. Brother David is quoted as saying that in a few years, with increase in faith, there will be no more material operations, only spiritual.

Among the many excellent color slides Mr. Voeks showed us were a number of his teacher Brother David Oligane. As a "medium," David calls on his "spirit protectors," St. Gabriel and St. Michael. Sister Juanita at the same chapel calls on St. Michael also. For four and a half months, Mr. Voeks went daily on healing missions with Brother David. David, David's sister, and several others are "active praying members" of this chapel, and these eight or ten people all live in a compound of several houses, where Mr. Voeks and Dr. Kudaka, a Japanese medical man, also stayed during their training. During this time, another Chicago woman was treated for multiple sclerosis, with a marked improvement, though to Brother David's sorrow, she was forced to leave after only five days because of her expired visa. Mr. Voeks' mother was also treated at the chapel for diabetes by Brother Mercado, with what our speaker described as completely successful results. He said this was doubly confirmed when the customary medication became unnecessary, and later when medical tests were negative.

The Espiritistas believe that God is the source of their healing power, as they take literally the Old and New Testaments. They make all their moves at the direction of the Spirit, and also believe that faith is required of the patient as well. Brother David said, "After you have healed by the Spirit it is very easy to heal the flesh."

In many of our speaker's color slides were seen the "banners" of the individual chapels and healers. These are "designed by the Spirit" and executed by a skilled person. They contain religious symbols and the healers, relying on the words of the Bible, believe that their powers comes to them through these banners.

Just two of the many things done by Brother David which would require you to "be there to believe it" are: he holds his finger six or eight inches above the patients' body, moves his finger quickly back and the skin parts and opens up. Mr. Voeks said that Brother David could do the same thing by taking hold of someone else's finger and using it in the same way, and that he himself had experienced it.

Brother David, while in trance, had produced, as one of the "nine gifts of the Spirit," tiny roses and orchids, less than half an inch across, while touching his banner. These were for presents for the two women patients from Chicago. After recounting in vivid detail a great many other equally amazing occurrances, he said, "Every day with David I saw incredible things."

The editor of this bulletin feels inadequate to the task of justly conveying an impression of what Mr. Voeks had to tell us, and of his fine, fresh spirit. Perhaps his comments on Tony's difficulties is as good a way as any. He said, "I think what's happened to Tony is our fault, not his . . . it isn't my place to judge him." It is also appropriate to quote from David Oligane's recorded words, which Mr. Voeks played for us: "This sick world of ours needs help from above." Thank you, Doug Voeks, for a most rewarding evening!

PHILIPPINE SITUATION: Comments from ARF: All of this newsletter, up to this point, has been read and agreed to by Doug Voeks himself, and he has also given me his address at the home of David Oligane, where he is now returning so that those who wish to write or visit can contact him through ARF.

Since Doug was there, martial law has been instituted in the Philippines, and napalm is being used to try to stop people who are opposing the govern-

ment of President Marcos.

Apparently, another Vietnam is starting right in the island of Luzon, where miracles are taking place and warnings of a Third World War are being clearly spoken. In the words of apostolic times, one ought to ask, "Brothers and sister, what shall we do?"

Philippine healers are not the only ones who have seen and reported fearful visions of the coming of World War III.

But others, with equally high credentials in things of the Spirit have warned us NOT to think and visualize evil things, for that is something that causes them to happen: "That which you fear the most shall befall you." Worry and fear are negative prayer—a way of visualizing and praying for the opposite of what we should envision.

Perhaps we should not neglect the sincere and loving warnings that have been given, but it may be equally foolish to dwell on such visions.

I cannot believe that the Spirit would give us visions of evil to come and then offer no alternative. It may be that for new wars to be prevented, a tremendous miracle is required, but the very work that those prophets are doing gives us proof that tremendous miracles are possible.

The chief requirement may be that we have to want that miracle so much that we see it as the very most important thing and start working at changing our lives in such a way that we become effective miracle workers, just as those others have had to change their lives to become effective psychic surgeons. That may sound hard, but it is not as hard as it sounds because it can be done by starting with one easy first step. And from that moment on, each step leads to exciting adventures which completely obliterate the boredom with which most people in our society are plagued throughout their lives.

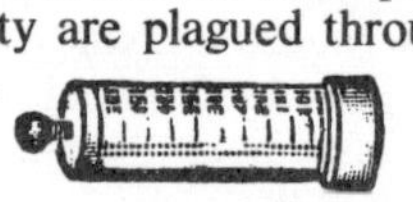

NATURAL BIRTH CONTROL AND MENTAL CONTROL OF CONCEPTION

The new book on natural birth control is now being worked on and we hope to have it in a few weeks. In addition to more charts for the astrological method and a section on predetermination of sex, the new book will contain more information on mental control of conception. We now have detailed information on the Murias of India, in which it seems that mental control is used almost exclusively by the young people who have complete freedom of sexual activity from childhood until marriage at 17 or 18, and only 4% of the girls become pregnant before marriage. Information on this was found in the book, **KINGDOM OF THE YOUNG** by priest/anthropologist, Verrier Elwin ($8.50, Oxford Press, Fairlawn, New Jersey 07410). This led to the possibility of research in mental control of conception, so that we are now planning an experimental class in this method of birth control.

The class will be taught by one of the area's best hypnotherapists, under the medical supervision of a top notch psychiatrist. In order to get started in the best way, we would like to meet and interview some of the people who have already been using mental control of conception for at least a year and who have had good results with it.

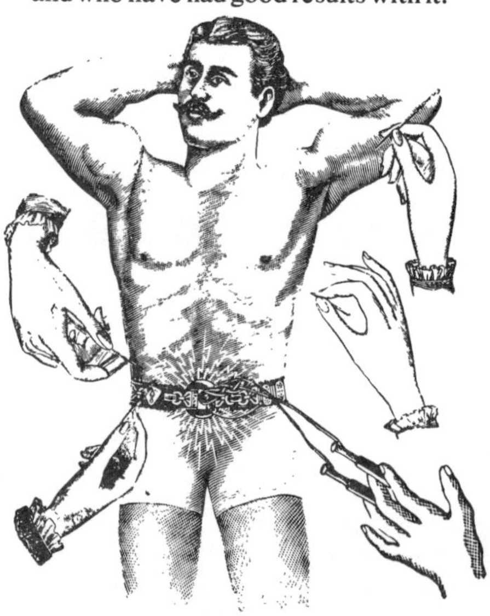

AURAS & ACUPUNCTURE CONFERENCE; the second in this hemisphere, was held in New York City, on February 13, 1973. A large number of speakers participated, slides and movies were shown, and a tremendous amount of information was exchanged. After a talk by a Western M.D., in which acupuncture was explained away as being mostly in the mind, a movie was shown by a California research group, showing a difficult dental procedure during which no anesthetic was used (except for two or four electrically stimulated needles placed in the patient's legs). Various areas of the patient's mouth were tested for anesthesia, and it was discovered that feeling was only numbed in certain areas where the work required it, but that other areas not far away were still quite sensible to pain.

KIRLIAN PHOTOGRAPHY was also a subject of the February 13th conference. There were actual demonstrations of this, together with many colored slides and detailed discussion of the process. One of the researchers in this field who worked at the University of Washington in Seattle is Richard Miller, and in a private conversation, he told of a simple experiment which seemed to prove that there is a simple method of TRANSMUTATION of ELEMENTS that can be repeated by any researcher. The experiment was described as follows:

CARBON can be changed to IRON. A copper dish should be coated with lampblack (pure carbon) by holding it over a candle until a thick carbon deposit is formed. Then a moderately strong electric arc should be played on the carbon with a carbon rod. After about 15 minutes of this procedure, the finely powdered carbon and whatever else was there should be shaken off on to a piece of paper, and a magnet should be moved around beneath it. At that point particles of iron should be drawn out of the carbon powder by the magnet and should become visible on the paper. People wishing to repeat this experiment should contact us for Richard's address and phone number.

PYRAMID EXPERIMENTS were conducted here by Jerry Boydston, but we have not yet had an objective proof of the effects of the pyramid. Meat did dry out in the pyramid as expected, but it also dehydrated in a hexagonal shape. Perhaps larger quantities of meat without preservatives need to be tried. Jerry has been building some additional pyramids, including one large enough to sleep in, and further experimentation will be undertaken. We telephoned Lynn Schroeder, and she gave us names of other researchers to whom Jerry has written. One of the people doing research in pyramids is the son of Marshall McLuhan, and Jerry hopes to hear from him. Jerry has also been working with a friend who has chart recording equipment, and they have found some unusual effects in chart recording the resistance changes in meat while it dehydrates. They are hoping to use a two channel recorder so that a control outside the pyramid can also be observed.

Jerry Boydston asks that anyone who is doing research in this field or who has done research in psychotronic generators should contact him by telephone at (215) 848-5497 or write to him at 4940 Germantown Avenue, Philadelphia, Pennsylvania 19144.

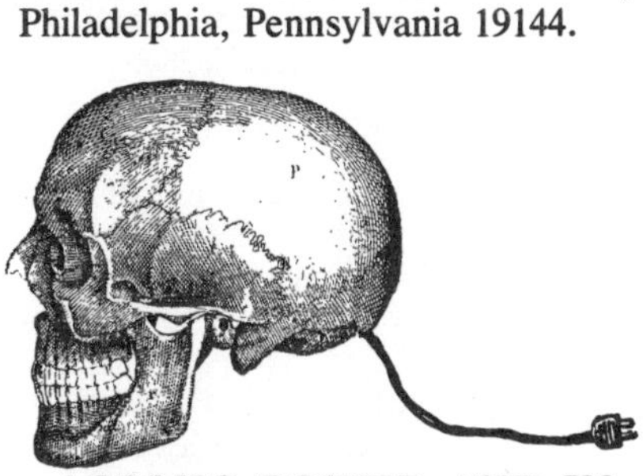

ATOMIC POWER AND INFANT DEATH: The following article on the effects of pollution due to atomic power systems seems to be well verified by articles and other sources. Seems evident that development of low cost solar energy is an urgent consideration. The materials to make solar cells are extremely plentiful, but very little money has been invested in research. Two places where research is being done are: U.S. government Jet Propulsion Laboratory and Arizona State University, where Dr. Charles Backus is a specialist in direct energy converters. I would suggest that interested people contact them, and I would be grateful to learn the results of such contacts. See article below for effect of atomic power production.

NUCLEAR REACTORS: DRINK DEATH, PITTSBURGH

By John Covert of Alternative Features Service

(AFS) A strong connection between radioactive wastes from nuclear power plants and lethal affects upon neighboring population centers has been found by a University of Pittsburgh radiology professor. His findings are helping arouse public resistance to the construction of new reactors.

Dr. Ernst Sternglass, a specialist in the effects of low level radiation on the human body at the university's school of medicine, has recently released a study linking nuclear waste discharges with increased infant mortality rates and with serious adult diseases.

Sternglass created a good deal of controversy two years ago when he reported that as many as 400,000 American infants under the age of one year may have died as a result of the radiation released from nuclear testing in the United States by 1965. He showed that infants are many times more sensitive to radioactive wastes than adults—something apparently no one even considered during debates on "acceptable" radiation levels. But it wasn't only infants who were dying, Sternglass said. The wastes associated with nuclear testing seemed to be causing large increases in the rates of respiratory diseases, cancer and leukemia among adults as well. This study was reprinted in part by **Esquire** magazine.

Predictably, that report brought charges of distortion and use of misleading figures from those who had a stake in nuclear testing. Protests came not only from scientists allied in body or in mind with the defense industry, but also from government agencies like the Atomic Energy Commission (AEC) which is supposed to be protecting the public from excessive radiation.

Since then Sternglass has been into a new project concerned with three wholly unexpected rises in infant mortality in sections of Pittsburgh and in nearby towns such as Aliquippa and McKeesport since 1960. Sternglass blames each of these rises on three excessive releases of radioactive wastes from nuclear power facilities around Pittsburgh.

On April 3, 1960, a serious accident occurred at Westinghouse Electric Corporation's Waltz Mills Materials Testing Reactor on the Youghiogeny River some 20 miles upstream from Pittsburgh. It was caused by an intense overheating of the radioactive core of the reactor, resulting in the creation of a lethal molten ash. The accident, unpublicized at the time for obvious reasons, released an estimated 5,000 curies of "highly radioactive fresh fission products" which found their way into the atmosphere and the Youghiogeny River.

Infant mortality rates in McKeesport had declined over the 1950's to a rate of 23.3 per 1,000 live births in 1956; but in 1960, the year the accident contaminated the McKeesport water supply, the rate jumped to 32.6 and in 1961, soared to 43.8, declining again in 1962 to 26.7 and in 1963 to 22.3.

Sternglass said that "a similar sharp peak in infant mortality was observed in the city of Aliquippa located some 30 miles further downstream from McKeesport, and for every county along the Ohio River downstream for a distance of some 150 miles."

Yet, throughout this whole period (1959-1962), there was an overall decline in infant mortality for the states of Pennsylvania and Ohio.

In 1965, there was a sudden rise in the gaseous and liquid waste releases coming from another Westinghouse facility, the Bettis Atomic Power Laboratories, located just outside of Pittsburgh in McKeesport. A good portion of the wastes—which at their worst never even exceeded the maximum permissible limit set by the AEC—ended up in the Monongahela at the point where McKeesport dips into it for its water.

In this case, too, Sternglass found another sudden jump in the infant mortality rate for McKeesport—a rise of 57 percent between 1964 and 1966—and again increases were seen downstream along the Ohio. This time the effects of the contamination were seen in Pittsburgh.

In the two years following 1968, Sternglass found a third rise in infant mortality that he laid to a renewed increase in gaseous emissions from the Bettis labs during that year.

It was easier to correlate infant mortality rates with increases in radioactive wastes because the impact is far greater and more quickly seen on small children than on adults, who can linger for years with cancer or leukemia before dying. Radiation can act on an infant's growth and metabolism, leading to immaturity at birth and lowered resistance to diseases.

But the ease of correlation with infant deaths doesn't mean there isn't any evidence linking radioactive gases released from nuclear reactors and nuclear tests with diseases in adults. In another paper, "Environmental Radiation and Human Health," Sternglass concluded that such gases "may have a serious effect on the incidence of chronic diseases of the re-

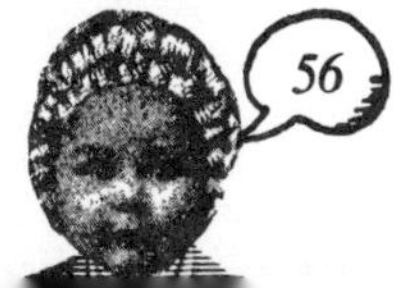

spiratory system such as bronchitis and emphysema that equal or even exceed the effects of conventional chemical air pollutants."

He found, for instance, that along with nuclear testing in New Mexico between 1945 and 1950, there was a sharp rise in death due to noninfectious respiratory diseases in that unpolluted state. Incredibly enough, that was **double** the death rate for the same diseases in the much more heavily industrialized state of New York. Prior to the advent of nuclear testing, the rate for New Mexico was very low.

Sternglass contends that the companies building new reactors are not only apparently unconcerned about the health hazards they pose, but that the companies are purposely scrimping on safety measures to save money. Westinghouse claims it can make equipment that will emit zero wastes, says Sternglass, "but it is, of course, expensive, and Duquesne Light (the building of the two Pittsburgh reactors) didn't opt for that.

UNDERGROUND MEDICINE

ARF was able to attend the 1st Annual Convention of the International Association of Cancer Victims and Friends, held on April 14th and 15th, 1973.

I managed to tape record almost all of the lectures with good quality reproduction, as I was able to connect my tape recorder directly on to the hotel's public address system, avoiding background noises.

Though the theme of the convention was cancer and natural cures for it, it soon became evident that cancer is only a symptom of poor health; that really healthy people don't get it—but the same is true of almost any other disease. Germs do not cause disease; the germs simply multiply when the body's defense system is too poor or weak to get rid of them, and then the person becomes ill.

Hence, the cure for cancer seems to be generally good nutrition and natural conditions-physically, emotionally, and spiritually.

Ann Wigmore of Hippocrates health Institute was there, and no speaker who was aware of nutritional matters seemed to be in fundamental disagreement with her therapy.

One exception to this was Dean Burk of the National Cancer Institute of the Department of Health, Education, and Welfare. Dr. Burk, who has been with NCI for 34 years and is head cytologist there, believed that Laetrile was the safest and best therapy for cancer, and he was quite upset with the men at the top of the NCI who used dishonesty to prevent it from being properly tested. He felt, however, that the ordinary researchers were only ignorant, not believing that anything which was non-poisonous in large doses could possibly be effective against cancer, and therefore, they did not wish to try it. He was the one person who did not know or believe that nutrition was an important factor. I believe that if he had listened to other doctors present, or to Dr. Krebs, he would have come to a different conclusion.

He reported, however, that in spite of the fact that it is illegal to use Laetrile (on a patient if you are a doctor), far more of it is used in this country than in the 22 countries where it is legal (and where there is probably less cancer). He reported that he frequently receives calls from medical doctors asking where they can obtain Laetrile, but in almost every case, the doctor wants it because he has cancer himself, or wishes to treat a close relative. Doctors do not use it to treat patients; to do so would, under present restrictions, they would risk their license to practice medicine.

Laetrile is now being synthesized by Dr. Ernst T. Krebs Jr., and it is being sold as a food (defatted kernels of apricot seeds) by a company organized by the man who set up the factory that makes it in Mexico. In this less refined form, it is less expensive than the $3 per gram from Mexico. The cost is approximately $1.40 per gram. However, it is in a form that can be eaten but not injected, and the amount that a person can assimilate in that way depends on the person's condition. I spoke with Dr. Krebs, and he agreed that it should be satisfactory in this food form, which is called, Aprikern. It is available from General Research Laboratories, 6925 Hayvenhurst Avenue, Van Nuys, California 91406.

MOST INTERESTING SPEAKERS: The two speakers that seemed to arouse the greatest interest were both in some way connected with the work of Dr. Schute who, I am told, was head of the Canadian Medical Association until he became a radical and represented the curative powers of vitamin E.

One of the speakers was a lawyer named John Joseph Matonis, and the other was a medical researcher working with Schute. His name is Dr. Rudolf Alsleben.

Both of these men publish their findings in a magazine called **Answer** magazine. Copies can be had from the Alsleben Medical Group, 710 N. Euclid Street, Suite 100, Anaheim, California 92801, Tel. (714) 776-9770.

Dr. Alsleben showed slides of what can actually be seen to be happening in human blood at a magnification of 1500X in a "dark field" microscope (which is one where the light is reflected off the specimen instead of shining through it—you see the specimen, not just a shadow).

Dr. Alsleben's talk was so dramatic and convincing that when he tried to leave, he was mobbed by people who wanted his address. He gave the crowd about 50 business cards and people just jumped all over each other to get them. I have to admit that I was in that mob, got a card, copied the information, and gave it away. I will send for the magazine and let you know about it when I have read more.

PRISON BREAK FOR HIGH SCHOOLERS !!! You can escape from high school and get an education, learning what you really want to learn! And you can get that dratted "diploma" in the process! That's what the Apprenticeship Service Program of Pacific High School can do for you! This program arranges for kids between 14 and 18 to leave home and "school" and go to live with someone who is able to teach them whatever it is they want to learn. While learning at the home and/or business of the teacher, the apprentice helps with the work he is interested in by actually working at it under the guidance of an experienced older teacher. Meanwhile, teacher and student send in written reports to Pacific High School, which fits the work into a suitable category and gives high school credit toward a diploma, which the student gets at eighteen. The arrangement is legal because the student is enrolled in a bona-fide high school program, and, while living away from home, is not bugged by local authorities trying to get their state money by keeping him locked up in a "school".

John Holt, author of **How Children Fail**, says of his program, "I think your proposal is one of the most constructive educational developments I've heard of for a very long time . . . and I think it could make a world of difference in the lives of many young people."

THE MENNINGER FOUNDATION, the country's leading psychiatric research institute, has tested some Silva Mind Control graduates, and have now become deeply involved in this work themselves. But they are using extremely sophisticated electronic equipment to make it possible for a person to be aware of his brain waves and other functions that used to be assumed to be entirely beyond voluntary control by ordinary people.

The ARF is about to acquire its own electronic Alpha brain wave feedback device, and when this is thoroughly tested, a report will also be given the readers of this newsletter. The unit being tested is designed to make it possible for a person to actually hear his own Alpha brain waves as they modulate a noise that sounds like a steady rain. The Alpha waves themselves are too slow to hear because they come at a frequency of 7.5 to 14 cycles per second. This instrument, which is battery operated and self-contained, allows you to hear the frequency, regularity, and strength of the waves, and thereby develop conscious control over them. It will sell for only $70. At present, I see such instruments as a possible help in learning to control the mind and body, but it seems that a course such as the Silva people have developed is also essential to understand all the possibilities, and realize that we are not dependent on electronic machines. The maker of this machine suggests reading the article by Collier in **Saturday Review** of April 10, 1971. It is called, "Brain Power—The Case for Bio-Feedback Training." (I have not read it yet.)

DREAMS TO REALITY! Aquarian Research started out with the proposition that we have to find much faster ways to overcome people's resistance to change.

Now, after almost two years of seeking, a breakthrough is coming. We have found a way that overcomes resistance to change among people of all kinds of background and now it has to be worked on and perfected—we need your help to do that.

I spent three days in Silver Springs, Maryland taking the advanced course offered by the Silva Mind Control people. We went fully into the history and development of this discovery which proves to all kinds of people that they are able to develop a genius mentality and deal with all sorts of problems far better than the computers and think tanks of big business.

CHILDREN learn to use the mind control method even faster than adults, and kids from eight years and up can take it.

EACH ONE who takes the course experiences for himself that clairvoyance is a real fact. At the end of the course, you are asked to bring in a card on which is written the name, address, age, and sex of someone with a serious illness. You are then able to test other graduates, and you find that they are able to visualize the person and sense just what is wrong with them, and they can do that with an accuracy that will amaze you. You will be tested yourself in a similar manner, and to your amazement you will find out that you can do it too!

Each one seems to go through the whole course thinking that others may be successful, but that it will not work with him—but it does. Then you begin to wonder if you shouldn't have listened more closely to all the other things that were being taught, but which you might have missed. It doesn't matter. You might have slept through half the course, and you will still be a winner in the end. Just to be sure that nothing you might have missed is permanently lost, you will then have the privilege of repeating any part or all of the course with any instructor anywhere.

There is no charge once you are a graduate. Taking the course will also entitle you to take it again for free, regardless of what the regular price would have been.

GRANDMOTHERS are also very welcome in the Silva Mind Control course. My own mother, who is over 70 (and an artist also), is planning to take the course. My dad, who is getting on to 80, wants to take it also, but does not feel ready yet, as he is still recovering from a recent heart attack.

I expect that he will take it later, though, as the course teaches one to relax so completely that the liklihood of further heart problems will be greatly diminished.

NEW AGE AND SOCIAL CHANGE EFFECTS OF MIND CONTROL: Most of the Mind Control instructors that I have met so far have come out of very conservative backgrounds. Two of them were engineers involved in sales. Another was a minister or priest. Yet all of them are now deeply involved with New Age thinking.

Most of the people who have taken the Silva course thus far are also from rather conservative backgrounds. They have to be to have the money that it costs to take it. Many are businessmen who had no idea of New Age thinking. They took the course because they felt it would enhance business ability (and it does). But now, they are beginning to see beyond that. Their minds have definitely been opened at least as fully as several good LSD trips might have done.

Each class is composed of people of all types (including blacks and nuns, hip and conservative). Everyone finds that they have the same abilities, regardless of where they are coming from.

FINALLY, however, from the time a person becomes aware of his own clairvoyant abilities, the notion that anything is "impossible" fades rapidly away into the past. In fact, miraculous medical healings and the miracles of the Bible begin to become obvious, and even a common experience among some of the graduates. Each time a medical reading is done, the student is encouraged to use his newly discovered powers to send healing to the sick person. Often, there are striking results.

The strange thing about the Silva method is that it has not become more widely known. Why is this? If someone has really found a way to train anyone to be a psychic, that is a tremendous breakthrough that ought to make headlines! That it doesn't, and the fact that important people in the field of psychic research still seem to be unaware that psychics can be trained, makes one wonder.

Here is my own experience: The first I heard about it was when a friend in Jersey City, who sends me clippings, sent me one of the advertisements for a seminar on Alpha brain programming. It looked to me like a very commercial, money-oriented thing that was promising much more than could possibly be real; like another one of those "success" courses for businessmen. And it cost $5 to attend. I wasn't going to bite. It couldn't be anything real or scientific if it advertised like that. If it was being held in the Bucks County Seminar House, I might also not have gone because they are so expensive and professional, but I would have believed it to be real. (But they would not advertise that way.)

Other people also sent me ads of the same seminar held locally, but I was not to be taken in.

Then the editor of **Thursday's Drummer** called me and asked if an old friend of his could stay at my house for a week while he took the course. I had not known until then that it was a whole week's course. As I read the ad, it only seemed to refer to one evening meeting, and the promises made were so extravagant that it couldn't happen that quick (if at all).

Of course, I invited his friend, and had him tape record the lecture for me (I wasn't going to shell out $5 for some gyp thing).

I was too busy to listen to the tape right away, but as the course continued, and he told me what was going on, it got more and more interesting. Finally I listened to the lecture and found it too religious: Bible quotes—"the Kingdom is within you" sort of thing.

But then this friend, David Miller, began to give me problem-solving techniques that he was learning; it seemed interesting, so I decided that I would like to take the course myself **if** they were sufficiently interested in the work of ARF to offer me a scholarship. I asked the teacher, and gave him one of our pamphlets. He told me to write him a letter and he would let me know.

I wrote and then waited a long time. I almost scheduled other things in the time when the course would be here, but then the acceptance came. I really didn't know if I should take the time, but when they agreed to my taking it on a scholarship basis because of the work I was doing, I guessed that they might not be so money-oriented after all, so I decided to take it.

When I did actually take the course, I was really quite skeptical. Also, I found it very difficult to keep awake during many of the conditioning exercises. I felt sure that others might manage to do clairvoyant readings, but I didn't think I could. And if I failed, I wouldn't be able to get any money back either! So at the end, I tried very hard.

It didn't work—I couldn't see anything at all. The instructor came over to help me. Didn't work. His impression was that I was trying much too hard! I really wanted it to work! Other people were getting it, though, and so I began to expect that I would also; then, indeed it began to work; the trouble was that I was trying much too hard. The others apparently were not trying so hard because they figured they would get their money back if they did not do it, but I hadn't paid!

I was able to do several accurate readings, and I got five people to do a reading on my brother in California. I had been talking to him on a mind-controlled telephone (free, of course) and he had told me to find out what was wrong with him. That was a "blind" test. No one could get it from me by "telepathy," so if they got it right it had to be directly from him. Four out of five found that he had trouble in the lungs and colon. So I called him again and gave the message. Being still very skeptical, he said, "OK, I'll give you 25%; I had a cold."

To face the world in eager wonderment grasping the promise of life.
To sense the life behind each gathering cloud and guess the meaning of the trees and grass
Yours is the vision, yours the prophecy—
Stretch yearning arms towards eternity and mold the dream into reality.

—John R. Brown

SUN BEAR VISITS ARF: The most important thing to happen at ARF this month was the visit of Sun Bear, founder of the Bear Tribe of American Indians, and his sister and co-worker , Morning Star.

Sun Bear, a medicine man of the Chipewa Tribe, farmed in northern Minnesota until he was 20 (Indians mature young), and then went on to do many other things, including teaching at a university in California (although he had only been through the eighth grade himself). He taught a course in survival, and has written several books (**Buffalo Hearts** and **At Home in the Wilderness**, from an Indian's point of view) on the subject.

As a result of his survival course, some were moved to try living on the land, and he went along with the idea. A new Indian tribe was started. By following the insights that came, they arrived at a new understanding. Eventually they came to see the old Indian prophecies of the coming of a white man who would devastate their land to have been fulfilled, and that it was now time for the fulfillment of the next prophecy: that the sons and daughters of the invaders would give up their "chrome" dream and become part of the Indian peoples.

For that reason, a new tribe was founded in the traditional way—for the first time in almost a century. The purpose of the Bear Tribe was to be a "medicine" (or spiritual) tribe which would begin to fulfill the old prophecies. Their door would be open to people of all races who wished to come and live in the Indian life—way as keepers of the land.

ARF REFLECTION

For a long time I found it very difficult that so many people who came to help at Aquarian did not decide to make this their home, and I felt that that represented a personal failure on my part.

Then I remembered what Kierkegaard had said about depression in **Either/Or Part II.**

"Depression comes when the spirit is trying to change something in a person's life and that person is resisting the change."

In the past, I had found that that was true and that if ever I felt depressed, I only had to ask myself what "change the spirit was trying to make in my life" or "what change was I resisting?"; and as soon as I was ready to let the change occur, the depression was gone completely.

So now I asked myself what change might be required and came to see that ARF need never really become a commune. Instead, it might be seen as a kind of free school on a graduate level where people just come for a time to learn what the next step for their lives might be. That was, perhaps, why we had to know about the Bear Tribe and all the other communal groups that were of some effect in helping the New Age to come.

So I gave up the idea of starting a commune here, and stopped trying to get anyone to stay on a permanent basis. In speaking later with some who had left, I learned that my pressing them to stay had also been one of the reasons they left.

A friend of ours named Steve Schwartzman has recently brought us into a personal relationship with a psychic named Ray Stanford, whom he knows, and was able to tell us that Ray is eager to help in the cause of world peace, which he sees to be the most urgent project at this time.

Over two years ago, when the idea of the ARF was first conceived, I had read of Edgar Cayce, who was able to answer questions with remarkable accuracy through the use of spirit communications or clairvoyance. I felt an urgent need to find people presently at hand who could do similar things, and who could help us to find out how to overcome society's resistance to change. Of course, advice from those other sources still has to be tested in a practical and scientific way, but if it is possible to get accurate advice quickly from such sources we ought to try to do that. The article that follows is written by Steve and tells more about Ray's work.

We will try as soon as possible to take up a contact with Ray and see if he is able to give us help in our work.

RAY STANFORD

In recent years, many people have learned about Edgar Cayce, the American psychic who devoted much of his life to helping the sick in body and mind. In the years since his death (1945) we've had time to study the great legacy he left behind. Now a new opportunity presents itself in the person of Ray Stanford, age 33, of Austin, Texas. People who knew Edgar Cayce personally say that the similarity between the two psychics—even the intonation of their voices when in trance—is striking.

Recently, Ray Stanford was in New York for a series of public lectures. At this time he has felt the need to make known the information that has come to him from a psychic "source" over the years. Ray's most impassioned plea is for peace, since he feels we are entering a period of crisis unlike any the world has ever confronted before. He asked people to pray as earnestly as they know how, and he offered the assurance that as few as eight people, if devoted, could forestall a war.

Ray is a psychic, both consciously and unconsciously. In the waking state he can see auras (which he tries to ignore since most people's auras, he says, are so dirty). He can also remember past lives, and he recalls doing that even at the age of five months; he describes seeing them as if they were a string of beads, and that vision led him to reject established ideas of a permanent heaven and hell.

In the trance state, Ray is completely unconscious and can later remember nothing of what transpires, though during these periods of trance, the psychic source reveals valuable information. His work is similar to Cayce's except that he does not accept physical readings; the basic reading he offers is what is known as a "self-help" reading, and corresponds to what Cayce called a "life reading." In these readings, a person's problems and aptitudes are dealt with in terms of past life experiences, but the emphasis is always placed on achieving a greater spirituality through the situations that come up in the present life. Even in this time of his relative obscurity, the waiting list for self-help readings is over a year long.

Because of the increasing requests for his help, Ray discourages what he calls "psychic shoppers." That is, people who only want information to satisfy their curiosity rather than to change their lives. He also warns people against pursuing psychic abilities for their own sake, and, like Cayce before him, counsels people to live as well as they know how and accept any resulting psychic phenomena as an extra spiritual gift.

In keeping with this idea, Stanford would prefer to give more "research readings" which would deal with subject areas instead of specific people. He feels he could be of greater benefit to everyone that way.

Ray's organization is the Association for the Understanding of Man, or

AUM. AUM's function is to schedule the readings, conduct them, and send tape recordings to the people who initiated the request. In addition, the organization has begun to publish a monthly newsletter and quarterly journal for members. The journal is especially valuable as it contains some of the most important information received from the Source. The first issue, which should be out now, contains the entire text of a reading on Fatima.

Apparently the Fatima vision has reappeared to various people periodically, and is most recently connected with a continuing series of visions seen first in a suburb of Cairo, Egypt, and presently in Lebanon.

Ray showed photos of an elongated glow in the sky just above a certain church; inside of the glow was what he called the "Angel of Mary," and overhead was a bird that seemed to hover without flapping its wings. Ray said that the vision appeared for a little while on each side of the church so that everyone in the crowd that gathered there each night could see it. He described the figure's expression as one of calm but grave concern, and said that she carried an olive branch in her hand, symbolizing peace.

The psychic readings seem to explain this vision (which Ray said has been seen by many thousands of people) as an urgent plea for peace. The vision warns against an impending Papal assassination and a world war that would involve the United States and Russia on one side against Oriental peoples on the other.

Also available from AUM are two publications. "Creation" is a booklet that explains the spiritual origins of man as revealed by the "Source." **The Spirit Unto the Churches** is a series of readings on the endocrine glands which, according to Stanford's source (and the Cayce readings), act as a connecting link between man's physical and spiritual nature. This is a profound book, meant to be studied, assimilated, and above all, applied. A free brochure about these books, the readings, membership in the AUM organization, subscriptions to the newsletter, etc. can be had by writing to: AUM, P.O. Box 5310, Austin, Texas 78703.

THE CALIFORNIA TIME Together with four others, I left for California in an old Plymouth on the morning of August 14, arriving almost without incident on the afternoon of August 16. We stayed overnight at a house in Davis which was being used as a base for the Bear Tribe Commune, and left the next day for a larger base known as "The Orchard" in Vacaville. That was the main Bear Tribe base for the reception of new people. On arrival we learned that Sun Bear had not yet returned from the East, but was expected back any day. He arrived two days later.

Some of the people told us that some sort of split had developed in Sun Bear's absence, and that, in general, things had become much more anarchical. The anarchy was always there. A great deal of useful work waited to be done, but not very much time was spent in doing it. People worked when they pleased and felt pretty free. Whatever was available to eat was shared by all.

The land belonged to a poor farmer who could not pay to have his peaches picked, so he offered the Bear Tribe half of whatever they picked. The peaches were cut and dried in the sun, and the Bear Tribe kept a supply for the winter.

Two things that we found about the Bear Tribe commune were very impressive. People would listen to one another, and people had a powerful faith and confidence that the necessary work would be accomplished, and they would help each other through whatever difficulties arose. The life style was extremely primitive, but we slept in the open and lived free as brothers and sisters, and loved one another in a beautiful way.

Pedro was so attracted to the life that after the first few days, he decided to stay, and when I came back two weeks later, he was still there and had no complaints.

PHOTOGRAPHING THE AURA After staying less than a week at the "Orchard," I went to visit my brother in Santa Cruz for a few days. He is a physics professor at the University of California there and also acting provost of one of the colleges. He was interested in—but skeptical of—many of the things I told him. In Santa Cruz, I was also able to find the Director of Mind Control instruction for the State of California, and found a very good contact with him. My brother agreed to take the course when it is taught in Santa Cruz, and I was given a promise that scholarships would be given to people who needed to take it but were, for good reason, unable to pay.

John Ille, the Silva instructor, also introduced me to Dr. Richard Brown in Los Angeles, a psychiatrist who uses the Silva method in treating the mentally ill. Dr. Brown is about to become a Silva instructor so he can also teach the method to his patients in the state mental hospital. His wife is taking the instructor's training at the same time. John also told me that José Silva would be quite interested in certain of the ARF projects, and I hope to share those with José when he is here in a few months.

Los Angeles was the next stop, where I was able to visit two engineers who were experimenting with the Russian methods of photographing the aura. One of them was Russian himself, and a Silva graduate also, and he gave me a copy of the circuit for the machine with which auras are photographed. He also translated the Russian text of the patent.

The other engineer was a friend of Dr. Karagulla, and he was doing the practical work of building and testing the machine. I was given one of the early aura photos of a leaf which he had made, and he explained to me some of the problems he had encountered in the design of the equipment. The work is proceeding very slowly because both of these men are very short of funds.

ONE WORLD FAMILY COMMUNE: On my return to Berkeley, Steve Scheffey took me to a large commune in Berkeley known as the One World Family. Here about 60 people lived together in two large houses under the leadership of Alan Noonan, who is believed by some of them to be a cosmic messenger who has been given his task by the space people of the flying saucers. Their aim is to provide abundance and peace to the entire human family, that we might live together as one world. The commune was well run by about ten of the people who were really dedicated to the ideals which Alan espoused, and who seemed to believe his claim. While there, I was able to talk with Alan (who really is a great artist), and he had me photograph sections of the sky where he claimed that, upon development, the film would demonstrate the presence of flying saucers. The film is not yet developed, but if the flying saucers (I saw none) do appear on the film, I will be taking Alan's claims more seriously.

The One World Family supports itself by running a restaurant in Berkeley, in which only vegetarian and organic foods are served. When I was there they were just about to open a second place. They also hold meetings in the restaurant to give their message to all who wish to listen.

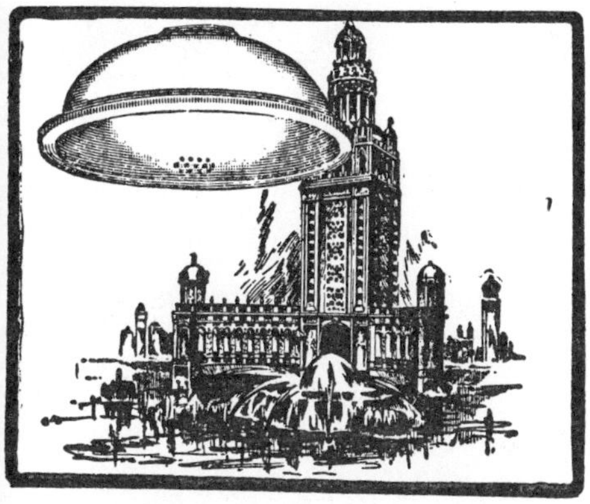

FLYING SAUCERS INVESTIGATED While in Berkeley, I was told of a man who had at one time been a contractor for NASA who had helped to design the Houston Space Center. I heard that he was trying to do a scientific investigation of flying saucer phenomena, and so I phoned him and told of my work. He was interested and came to see me the next day.

I learned that he also knew of the monuments on the moon which NASA had photographed, and he confirmed that Boeing had published the photos in an employee newsletter. He told me of a book called **Maps of the Anciet Sea Kings** by Hapgood, and said it proved that there existed accurate aerial maps of the entire globe long before Columbus' time.

JIM JONES & PEOPLE'S TEMPLE It was during one of these meetings at the restaurant that I met a man who told me of the work of Jim Jones at People's Temple.

He told me that Jim was a fantastic healer and clairvoyant, and that his group was concerned for a communal/socialist way of life which would be an answer for capitalism and for churchianity as well, and that he was a prophet and an apostle who could call me out of the audience and heal me or tell me things about myself which could not be known to him except by clairvoyant means.

That man, James M., agreed to take me to their place (a distance of 135 miles) and we rode up there with him on Sunday. Dianne, a member of the One World group, came with me also.

When we arrived, we found a large but simple meeting hall, the last quarter of which was taken up with a swimming pool (for the people). Soon the meeting started, and I was amazed to find that what James had said was true. The group was fully integrated with black people and white people in almost equal numbers.

Many testified with regard to Jim Jones' amazing power of clairoyance and healing. Finally, Jim came in himself and began to speak in a very warm and human way. First he called a bishop out of the audience and the bishop for California (of the Disciples of Christ Church) testified that this was the largest and most unusual of all Disciples' churches. He told us that he knew Jim well and that he could verify the things he was saying and doing.

Then Jim proceeded to explain why People's Temple was affiliated with a larger denominational group, in spite of the fact that the other churches were doing so little good.

He also read a testimony from a woman who is in prison for political crimes and because she is black, and told how he stood with her even though he did not agree with all she had done.

In a little while he began to call people out of the audience and tell them things about themselves which obviously surprised them. A few were told that their illnesses would be taken away and some were instantly healed, to their great amazement and joy.

It was hard to believe all this, and I could wonder if it was not a prearranged put-on job until he called on Dianne and gave her the name of a member of the commune in which Dianne lived (which was an unusual name) and told her how that person could be helped by him. Then he called me out of the audience and told me the exact names of all the people who were at that moment staying in our house in Philadelphia. When I checked on this by telephone, I found that he was exactly correct! (Even though I did not remember one of the names and at first denied it, and had given those names to no one at all outside the house.)

Later Jim spoke to me about my personal life, showing that he knew just what problems were on my mind, and though he did not promise to help with them, they did go a lot better after that.

Tremendous clairvoyant and healing power is only the small part of Jim's message. The great surprise was to find a man who had such powers, and at the same time was fully aware and concerned about the social and political situation in the country, and who was really causing people to get busy and do something about them.

Some of his followers believe that he never errs, and all that he does is correct. I am not of that opinion, but I am convinced that he is one of the most outstanding people I have ever met, and that he should be listened to with a great deal of attention and respect. Jim Jones can be contacted at the People's Temple, P.O. Box 214, Redwood Valley, California 95470.

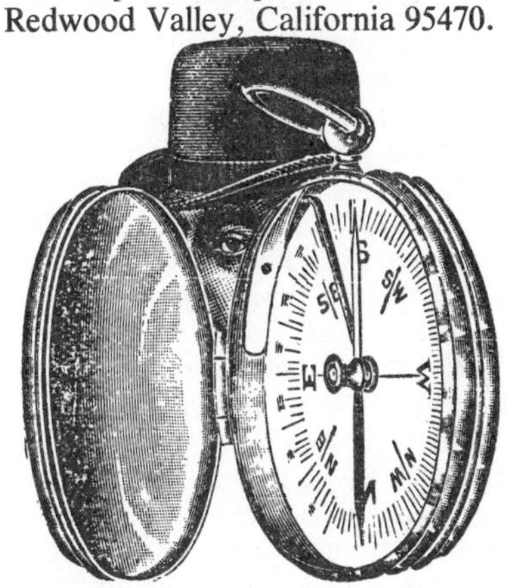

A PSYCHOKINESIS EXPERIMENT

A Philadelphia Mind Control graduate told us that she could cause a compass needle to move with mental power. To test this, I put together an instrument that used light from a flashlight bulb to illuminate the compass needle while a sharply focused photoresistive cell detected the movement. A movement of as little as one degree was able (when adjustments were correct) to cause the resistance of the photocell to change, and thereby vary the pitch of an oscillator that could be heard by everyone in the room, and also tape recorded. At the same time, the voices of the experimenters could be heard on the same tape.

In the short time that we were able to get everyone together to perform the experiment, we actually saw the compass needle move on two occasions, and one of these was recorded on the tape along with the voices of the experimenters. The experiment left little doubt that ordinary people were capable of psychokinesis of that kind, but that a great deal of time is required to produce the effect.

We also used the same equipment to check out a report of similar movement which we read in the book **New Worlds Beyond the Atom** by George De La Warr (out of print), but we found that the method which he described did not seem to work for us at the time.

JIM JONES IN PHILADELPHIA

Jim Jones and the People's Temple of California recently brought a bus load of people all the way across the country to hold meetings in Philadelphia.

Perhaps it was something other than "coincidence" which led me to visit the People's Temple group in California just a few weeks before they had planned to come to Philadelphia. I expect so. According to Jim, "Things don't just happen, they happen just!"

Jim and about fifty of his people came as scheduled, and the meetings were held at the Benjamin Franklin High School, Malcolm X Hall. Most of the audience was composed of poor people and blacks, and a number of our friends were also present.

Jim Jones gave the impression of being a very gifted performer, and, at the same time, a person with very great love for the poor. Some healings of serious illness appeared to take place, but have not yet been verified, as such investigation was not encouraged by the group and would have been awkward. A number of people were called out of the audience by name, and told things about themselves. This seemed to be a powerful demonstration of clairvoyant ability, and many were moved by the experience. However, some people to whom psychic experience is not foreign had doubts, and it was pointed out that most of what was revealed might have been obtained by other methods.

Far more important and moving than any psychic demonstrations was the love which Jim showed to many people, old and young, who needed help. Many people, especially those who were old or poor, were invited to come to California and live with the group at People's Temple. They were promised free transportation all across the country on the Temple's own Greyhound buses, and that when they arrived there, all their needs would be taken care of. As far as we know, this offer was sincere and was well kept for those who were prepared to leave with the group on Saturday evening.

As for healings, there was only one person who was said to have been healed that we have been able to check on, and in this case, a considerable improvement is reported, but it will be at least another few days before a definite word can be said about that case.

Some of us had reason to doubt the displays of clairvoyance that were given, and I talked with Jim about this. His stand was that he is first and foremost a socialist and that his purpose was to bring communal living socialism to America (and especially the poor), that he did not consider demonstrations of healing and clairvoyance to be of any importance except in so far as they are a tool for socialism. He stated clearly that he believed that his purpose was a good one, and that a good end justifies the means that may be required to achieve it. I find myself unable to agree with Jim on this matter.

Government plans for black people were also of great concern to Jim. He mentioned a document known as the "King Alfred Plan," which was supposedly sent out to various types of police agencies all over America. The plan describes in detail how the different agencies should deal with an uprising of American blacks if such should occur in a sustained and coordinated manner, and it was reminiscent of Nazi Germany in every way.

We found a copy of this plan and showed it to an attorney who is a close personal friend of Jim Garrison, the district attorney of New Orleans who had tried to prove the conspiratorial nature of John F. Kennedy's assassination. This man was able to show us good evidence that the "document" is not a government document at all, as it did not bear any names of people by whom it was supposedly sent out.

Later, we were informed that the "King Alfred Plan" originally appeared in a fictional account of a future event—a novel, and not an actual record.

However, we also discovered the existence of another document that actually was printed by the U. S. government printing office but which, we are told, is now out of print. This document is genuine and bears the names of the people who published it. It is an anthropological study of how black people can be confined to ghetto areas, and, like rats in overcrowded conditions, they would be emotionally and psychologically incapacitated by those conditions. It also refers to the killing of large numbers of blacks in Vietnam as intentional. We do not know to what extent this document is responsible for present government policy, but we know it is genuine and unloving in the extreme. We do not yet have a copy of it.

THE TRAVELER'S DIRECTORY IS OUR PASSPORT TO THE WORLD and a solution to the inflated cost of traveling. With persistent effort and a little help from our friends, it is finally to be reissued in not one, but two very much enlarged editions.

The Traveler's Directory is a kind of passport that enables travelers from all over the world to exchange hospitality in their homes, and thereby help each other to find friends wherever they may be, and to avoid paying for accommodations in places where they would otherwise not know a soul.

According to the plan, each member of the directory agrees to offer hospitality to other members whenever it is convenient, and in return receives the same from them when he is away from home. No one (according to the original plan) can receive a copy of the directory except by being listed in it. Thereby, whenever someone asks to meet another listee, it is possible to look up that person in the directory, and read the description of himself that that person has previously submitted to the directory for publication.

Each listee was permitted to submit up to 100 words about himself, and the kind of help or hospitality he is able to offer to other travelers. In 1969 (the last issue), there were about 300 people listed. In the planned final edition for this year (1973), we may have as many as 4000. Here at ARF, it was felt that the Traveler's Directory could be a way of getting a lot of different people into contact with one another, which could also help to open people's minds to the possibilities of the New Age to come. Therefore, we decided to tell all about it.

Month after month, we worked on the publisher and sent out letters to old members until finally, David Miller, who had been responsible for the work once before, was able to prevail upon the publisher to let him have all the material that had accumulated over the years, and a portion of the funds, adn we are now about to let you all know that it is a reality once again! The address is 535 Church Street, Lancaster, Pennsylvania 17062 (Tel: (717) 397-3472) Membership cost is $5 per year.

PSYCHIC RESEARCH AT ASPR in New York City is moving along. After visiting with José, I went to see Karliss Ossis of the American Society for Psychic Research. It is likely that they will receive a large sum of money from the will of James Kidd, who left his money for the use of people who will do scientific research to prove the existence of a human soul. Dr. Ossis told me that he is thinking of investigating "astral travel," or the "out of the body experience" as one way of proving the existence of the soul.

In an effort to learn more about that, I have been reading the book **Journeys Out of the Body** by Robert Monroe. It seems to be quite a scientific account and has an introduction by Charles Tart, who edited the book **Altered States of Consciousness**. Tart is a professor at the University of California at Davis.

BUILD A BETTER COCKROACH TRAP and see if the world will beat a footprint to your door. Perhaps, at any rate, this is what a cockroach told me about traps: roaches can climb vertical surfaces UNLESS those surfaces are smooth and greasy.

So take a large tin can or bottle with a wide mouth and glue a sheet of papers around the outside. Then prepare some bait made of a mixture of bacon grease and all the good stuff the little dears love best. Put the bait around the bottom inside corner of the container and rub a tiny amount on the outside paper also.

Next, rub some raw bacon fat all around the inside of the container (except the bottom)—just a thin layer —and rub a bit of raw fat on the outside paper too. That will make the inside surface greasy, and the outer surface will attract the roaches. As there is not enough to eat outside, they climb up the paper and go over the top to the inside. They slip—kerplump—right to the bottom and there they stay. Next A.M., you take the trap and flush the little dears down the toilet. Once a week, you have to clean it out and put in new bait and grease. We've gotten rid of hundreds of roaches that way and no kinds of poison are needed.

LADY HAS PUPS! Yes, seven little German shepherd pups, growing all the time, and almost ready for adoption. Two are already spoken for, and we expect that the rest will go quickly.

NEW BIRTH CONTROL METHOD SAFE, ACCEPTED. That is the title of the following article that we received from Puerto Rico a little while ago. The text of the entire article follows:

Los Angeles (AP)—The Roman Catholic Archdiocese of Los Angeles plans to set up clinics to teach a natural birth control method that a spokesman says is "safe, 100 percent sure, and morally acceptable to the Church."

Called the "ovulation method," it has "made the rhythm and temperature methods obsolete," the Reverend Robert Deegan, director of the archdiocese's health and hospital department, said in an interview.

The method was developed by an Australian husband-and-wife team, Drs. John and Lyn Billings. A book they've written about it sold more than 50,000 copies in Australia, and is soon to be published here by the Roman Catholic Church, Father Deegan said Wednesday.

The Billings say their method is based on the same principal as the rhythm method, but is much surer, and does not involve counting days from one menstrual cycle to the next.

They say the method is based on a discovery that a woman can become pregnant only on days when a mucus has been secreted by glands in her cervix.

"In every fertile cycle the woman notices this secretion of cervical mucus," says Dr. John Billings, "so it is an infallible method of determining when conception can occur."

The Billings are visiting the United States to lecture on their method. Last week they attended a conference near Washington D.C. held under the auspices of the Department of Health, Education and Welfare, and this week they are lecturing at Catholic hospitals throughout Los Angeles, where clinics will be established.

After receiving the above article, I sent an air letter to Reverend Deegan on March 1, but received no answer. This evening I phoned him (person to person) in order to find out if the article is factual. In the phone call, Reverend Deegan gave me the following information. The Method is being taught by volunteers, about half of whom are nurses. They also have an advisory committee of doctors. It is being taught at the Harbor General Hospital of the UCLA School of Medicine in Los Angeles.

The Billings were here one and a half years ago under the auspices of the Archdiocese, and it was arranged to find out what local doctors would think of the method. The method was known, of course, by doctors, but has never been researched much in this country. But the Billings have been using it for over 15 years in Australia, first in Melbourne, and now throughout Australia; the Archdiocese had arranged for the publication of an American edition of the Billings' book, and the first printing of 5,000 copies was sold out in one month.

Any woman should be able to use the method, which is based on the proper interpretation of a mucus discharge. The claim is that when understood and used correctly, the method is completely foolproof. Several of the nurses who are teaching it have been using the method themselves for several years.

DISADVANTAGES

The Billings' book was reviewed in the **Journal of Obstetrics and Gynecology of the British Commonwealth**. According to that review, the method has one serious drawback which Billings cannot overcome. He cannot find any substitute for the self-control required by the husband and wife's periodic abstentation. "Although this method of birth control is the cheapest known, self-control is very unfashionable these days."

As with the other method that ARF has publicized, there are brief periods of abstinence, and then periods when abstinence is not required.

PUBLICITY—According to Reverend Deegan, the method has been publicized by AP and picked up by various papers, especially in New York State.

In answer to my question, Reverend Deegan also said that there is more than one time during the months when abstinence is required, and he would have been happy to enlarge on that point, but for the fact that I was afraid to speak that long at over one dollar per minute. I will order a copy of the book.

BOOK PRESENTLY AVAILABLE

The book is presently available. It is called **The Billings Ovulation Method**, and may be had for $2.65. Order from: The Borromeo Guild, 1530 W. 9th Street, Los Angeles, California 90015.

OVULATION AND ASTROLOGY: Some months ago, E.R. Schweighart of Astra International in Vienna wrote to me to explain the relationship between the astrologically determined "cosmic fertility period" and "ovulation" as it was generally understood, and he reported that it was difficult to explain the cosmic fertility cycle in terms of the old rhythm or ovulation theories. He said what he did seem to understand was that a woman's cervix is sealed by mucus, which prevents the entry of sperm during most of the month, but that during three periods, this mucus disappeared and allowed the possibility of fertilization. These periods were during the "cosmic fertility time," during the rhythm fertility time," (or some part of it) and during the menstrual period.

With the information now being received from the Billings, it seems possible that we have a way of checking on the cosmic fertility period, and this would be worth investigating. The Billings' method may also provide a way of shortening the time of abstinence required during the rhythm period. From the scientific point of view, it would be most interesting if the Billings' method could substantiate the cosmic period and allow us to learn how that period may vary (if it does), or how precise it is (if it is exact).

SOUL TRAVEL

In Syracuse, ARF attended a meeting of Eckankar (called "The Ancient Science of Soul Travel") and heard a lecture by Darwin Gross, who is considered their highest leader in this country. He succeeds Paul Twitchel, who has written many books on this field.

For me, the most impressive evidence of the value of their work was the paintings and other works of art which were displayed there, and the generally pleasant atmosphere of a large group of about 500 people of all types. The general idea of Eckankar seems to be a belief that by "soul travel," a person can come to know God and the "higher worlds." They claim that what is often referred to as "astral travel" (which Robert Monroe reports on in **Journeys out of the Body**) is only traveling into the lower, or "astral" plane but that their methods of spiritual development bring a person into the "higher realms" or the "true spiritual experiences." People were given warnings against the other methods and told not to seek spiritual development in anything but what they claim to be the true path.

Prevention The magazine for better health

DECEMBER, 1971

Laetrile–
An Answer to Cancer?

There have been numerous natural treatments for cancer and every one has been banned — untested. Not all could be of genuine value, but neither are they all necessarily worthless. Without unbiased investigation, neither the medical profession nor we will ever know.

LAETRILE, WHICH 30 YEARS AGO WAS A controversial anti-cancer agent made from apricot pits is today, in 1971, still a controversial anti-cancer agent made from apricot pits. Though it has been in and out of the headlines, hailed as nature's answer to cancer and condemned as a worthless sugar molecule, actually the average doctor knows no more about it today than he did 30 years ago.

And, according to a recent edict issued by a five-man panel of cancer experts convened by the Food and Drug Administration, the scientific community will continue to remain in the dark as far as Laetrile's effect on human cancer is concerned. Your physician should he desire to try this agent on some patient who asks for it and is sure to die on orthodox therapy, is forbidden to do so according to FDA's latest ruling (August, 1971).

Thus, in spite of the fact that the federal government and most politicians are now firmly committed to the establishment of a "Conquest of Cancer" agency — funded with hundreds of millions of dollars — it appears that not one dollar will be used to conduct Laetrile tests on humans, even though advocates of Laetrile include a number of outstanding doctors and scientists around the world — in Germany, Italy, the Philippines, Canada, Mexico, Belgium and a few in the United States.

To explain their rejection, the cancer specialists said that Laetrile might be poisonous. This is directly contrary to the results of extensive tests made by Dr. Dean Burk, chief cytologist of the National Cancer Institute, who has reported that Laetrile is "remarkably non-toxic . . . compared with virtually all cancer chemotherapeutic agents currently studied." It is also not much of an explanation in view of the fact that every substance used in the accepted treatment of cancer, drugs and radioactive materials alike, is definitely and beyond doubt poisonous.

We do not intend to take up the cudgels for or against the effectiveness of Laetrile. We are not in a position to say definitely, "this is the answer to cancer. This is what the world has long been waiting for."

But, we *are* going to take a stand on the right of freedom of choice when it comes to the healing arts. Particularly with regard to cancer, where orthodox therapy consistently fails, if a physician wishes to experiment with a new treatment he thinks may work better, and the patient is fully informed that he is being treated experimentally and wants it, we can see no reason whatsoever why the government should interfere and decree that the patient must die under the usual treatment.

Nobody associated with Laetrile is getting rich on it. It is not patented and even if it should come into general use, nobody ever could get rich on it. So there is little or no chance that anyone would ever promote Laetrile to extract money from gullible or desperate people. Its promoters *must* be sincere, since uniformly they could earn more money with less trouble by forgetting Laetrile. They could be mistaken, but they are certainly sincere.

This being so, we think it is criminal on the part of our government not only to refuse to test this treatment and deny it to those who desire it, but to actually try to forcefully prevent people dying of cancer from getting into another country where they can get it legally.

Considered Vitamin B_{17}

Laetrile is a natural substance — discovered by Dr. Ernst Krebs Sr. of San Francisco, California, who was always careful to explain that he did not develop Laetrile — nature did that. He considered it a vitamin, which he called B_{17}, maintaining that Laetrile is to cancer what vitamin C is to scurvy; what vitamin B complex is to pellagra; what B_{12} is to pernicious anemia. His son, Dr. Ernst Krebs Jr., says in a paper, *The Laetriles — Nitrilosides in the Prevention and Control of Cancer*, (published by The McNaughton Foundation) "Cancer is a chronic disease and no chronic disease has yet found therapeutic or prophylactic resolution except by accessory food factors (vitamins) which are normal to the normal diet; one of the truly fixed laws of physiology, without exception. Night-blindness, once considered incurable, found resolution in vitamin A; Beriberi through vitamin B_1; rickets through vitamin D. Even mental illness is now known to be due to the deficiency of micronutrients normally present in the brain."

For background on the history of the use of Laetrile, PREVENTION went to Andrew McNaughton, an energetic man who has completed a successful career in business (stocks and investments),

Ernst T. Krebs, Jr., a biochemist, is the leading investigator of the properties of Laetrile, discovered by his father.

Continued →

After the lecture, each one had a chance to greet Darwin Gross (at least with a handshake), and I asked him if he was familiar with José Silva, Robert Monroe, or Charles Tart (who is an authority in the field of altered states of consciousness). Mr. Gross was not familiar with any of these. Obviously, the whole matter requires further study, and we have been given the name of a person in the Philadelphia area who is a leader of the Eckankar group.

In Buffalo, we were able to have a very nice visit with Sister Justa Smith, who was the main person responsible for the founding of the Human Dimensions Institute. We were able to spend the whole morning with her, and she showed us her college's research laboratory where scientific research in the field of healing is being done.

In one very important experiment, a man by the name of Estebany was allowed to hold a sealed test tube containing enzymes for different lengths of time. Mr. Estebany was known to have had a healing effect on various people, and the test showed that the activity of the enzyme (trypsin) was increased by his holding it. Tests were also made with others who claimed to be able to heal. Those who claimed no such ability (college students) had no effect on the enzyme activity. Those who had made such claims, but who had not been proven, did have a positive effect at some times, but not regularly.

A very interesting discovery was that the use of a high powered magnetic field (8,000 to 13,000 Gauss) had a similar effect in increasing the enzyme activity, though it was not determined if this increase occurred in the same way in the physical or chemical sense.

As it happens, Charlie Moore is planning to stay with some people in New Jersey who have been working with healing by the use of high magnetic field equipment while he is attending the New Jersey Conference on parapsychology. The magnetic healing equipment is made by Japanese firms, and I hope to obtain more information about this field of work in the near future so that I can then report on it. Like the Chinese use of acupuncture, such methods are as yet almost unknown in this country, and this seems to be due mostly to the influence of certain medical organizations which operate in the manner of a labor union.

For further information on the work of the Human Dimensions Institute, it would be well to subscribe to their quarterly magazine called **Human Dimensions**. A subscription is only $2.00 per year, and the publication is extremely well done. The address is: Human Dimensions Institute, 4380 Main Street, Buffalo, New York 14226.

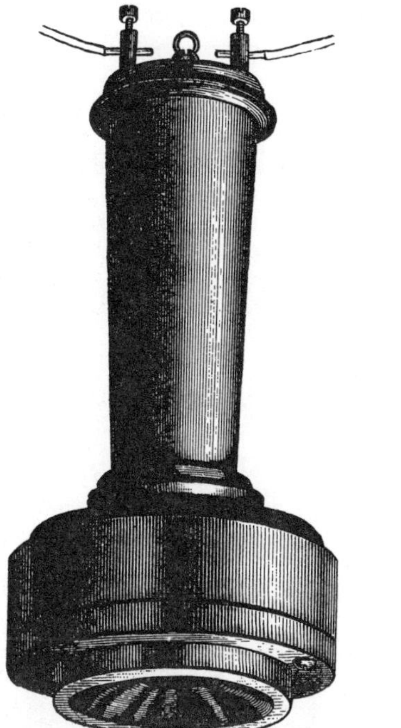

ZBS MEDIA—COMMUNE OF MIND CONTROL GRADUATES.

On leaving HDI, we drove steadily all across New York state. We arrived at about 6:00 P.M. and had a very warm reception by the ten or so people there, all of whom were Silva Mind Control graduates. In the short time we were able to stay (about six hours), we found them to be a very harmonious group who are aware of many things known only to a few.

ZBS has a house and a recording studio in a beautiful rural setting (where stars can even be seen—yes, real, natural stars). They record special radio programs, which are then sold for airing on a number of FM and other stations across the country. The only thing we had time to listen to was a "mystery" story, which is serialized under the title, "The Fourth Tower of Inverness." It was quite entertaining, and yet, not as mind-opening as actual discoveries now being made. However, it seemed like a good start to get people interested in a more open-minded view of possible reality.

Some of the things the ZBS people seemed to know about, also in the fields of healing and astral travel, and the whole style of their communal life, was so interesting that I would like to return there for a longer visit later on.

a former chief test pilot of the Royal Canadian Air Force, world traveler, — a man with immense vitality, perserverance and imagination. As head of the McNaughton foundation, a non-profit philanthropic organization, he has given extensive support to the testing and development of Laetrile. McNaughton answered all questions quickly and openly, even the adversary type query such as "What's in it for you if the government gives the go-ahead for testing?"

"I don't know or see how the Foundation will profit financially if Amygdalin (another name for Laetrile) gets into general use — any profit won't be through an effort of ours," he said. "By the end of 1972 I believe Laetrile will be in the hands of the establishment — and we'll be happily, and enthusiastically tackling some other problem. We have almost no interest in treating cancer patients; we are going this road only to get into the field of using it for prevention. As vitamin B_{17}, its prime function is prevention. *Amygdalin is in the public domain, no one owns it, any pharmacist can manufacture it.*"

The McNaughton Foundation is a non-profit organization formed originally in Canada. McNaughton, himself, has made substantial contributions to it over the years. Other sizable donations have come from the government of the province of Quebec, Timex Corporation and several other large industrial organizations in the United States, Canada and Europe, as well as public spirited individuals in other countries.

Serving with McNaughton (whose late father was General A. G. L. McNaughton, commander of the Canadian Armed Forces in World War II and president of the Security Council of the United Nations and of the National Research Council of Canada), are three directors and 13 advisory board members, all with distinguished credentials in the sciences or in the business community.

What is the purpose of the McNaughton Foundation? ". . . to encourage the creative scientific atmosphere in which the researcher follows his own line of investigaton toward the solution of the problem. even if these lines are in opposition to prevailing scientific or political opinion . . .

"The attention of the McNaughton Foundation is focused on transforming into reality new solutions to the problems of mankind. not on commercial developments *per se*, or ivory tower research, or educational goals. Thus, it is able to foster immediate development of promising solutions to human needs."

"The California Medical Association and the FDA term Laetrile therapy 'quackery,'" we pointed out, "what do you say to this?"

"I give you Dr. Burk's definition," McNaughton replied. "'A quack is a person who pretends to knowledge which he doesn't have.' We in Laetrile, if we were claiming things for Laetrile we don't know to be true, could, indeed, be called quacks. But, in fact, we are understating the real potential of Laetrile and the claims we make for it are based on its use in more than 5,000 patients.

"What do we claim for it? That in advanced cases of cancer it has a palliative effect, reducing the need for narcotics, improving the subjective condition of the patient — he feels better and doesn't have pain. Objectively, his blood pressure picture improves and he gains weight and frequently goes back to work until eventually he dies of cancer because he's an advanced case. But he dies without the pain and the total dependence on narcotics, generally because the damage to vital organs has eliminated his ability to sustain life.

"We claim it on the basis of thousands of cases observed. In earlier cases where there's not been damage to vital organs beyond the point where there's inadequate capacity for the organ to sustain life, in those cases patients frequently live to die of other causes.

"The real quacks are those who pretend to knowledge they don't have about Laetrile, and condemn it without knowledge. We have many patients," McNaughton said, "who say Laetrile has cured their cancer. We don't agree — we never say it cures. At best, we say it will eliminate a tumor. To remove the tumor, however, is only attacking symptoms. The disease is an underlying deficiency of nitrilosides (B_{17}), inadequate pancreatic function, the presence of trophoblasts at a site of estrogen production."

Organic Cyanide Considered Harmless

"Laetrile is an Amygdalin," McNaughton explained. "Amygdalin is a simple chemical compound consisting of two molecules of sugar, one molecule of benzaldehyde, one molecule of cyanide tightly bound together."

McNaughton smiles when he reflects on the furor created by efforts to win FDA permission to test on humans. "After all," he points out, "Amygdalin has been around for at least 3,000 years — the ancient Chinese used bitter almonds — with properties similiar to those in the apricot seed — as a medicine. But bitter almonds, by agreement with the U.S. Food and Drug Administration are not on the American market because one of the elements is cyanide — and FDA thinks cyanide will kill you."

According to McNaughton and the developers of Laetrile, cyanide naturally occurring in food is not dangerous. Primitive people who eat only whole natural organic and unfractionated foods, they say, regularly ingest from 250 to 500 milligrams of organic cyanide each day. This natural cyanide is locked in a sugar molecule. It is normal to our metabolism and is found in over 1200 known unrefined foods and grasses. When it is eaten and taken into normal cells, the enzyme rhodanese detoxifies the cyanide and releases it into the urine.

But, — and this is the rationale of Laetrile treatment — cancer cells are completely deficient in rhodanese, and are surrounded instead by another enzyme called beta-glucosidase. This enzyme, they claim, which is secreted by the cancer cell, in turn releases the bound cyanide from the Amygdalin at the site of malignancy, and the released cyanide destroys the cancer cells. Thus, in the belief of its advocates, organic cyanide is a highly selective substance which shoots at the enemy only. It is toxic to cancer cells — and completely non-toxic to normal cells.

Pearl Buck refers to glucosides in the treatment of cancer in ancient China (Laetrile is a glucoside). In 1834 a German chemist started chemical and pharmacological studies of Amygdalin. Dr. Dean Burk of NCI observes wryly, "This drug has been studied for 130 years, and more is known chemically and pharmacologically about Amygdalin than most drugs in general use."

Dr. Ernst T. Krebs, Sr., of San Francisco was the first in this country to experiment with its therapeutic properties. He worked with it from the 1920's till his death about 2 years ago. Another pioneer in its use is Charles Gurchot, Ph.D., former biochemistry instructor at Cornell University Medical School, author of *Biology*

Continued →

JIM JONES had his meeting in San Francisco on Sunday, and Saul came with me. Our meeting with him in San Francisco on this occasion was, nevertheless, a disappointment, as it did not seem that he was keeping to the spirit of the promises that he had made to me earlier. It is the purpose of our work to tell you about positive things that help bring about a New Age, rather than to expose people and groups where deceit is discovered. So, all that we need to say about Jim is that we are uncertain of his spiritual gifts, but he does appear to be a convinced socialist with many unusual gifts and originality. Perhaps he is a great help to many.

My next visit was with Robert Mattson of the Academy of Parapsychology and Medicine in Los Altos, California. I was impressed by the fact that this group is under the directorship of Elmer Green, who is also the director of research for the Menninger Foundation in Topeka, Kansas, and is one of the leading people in the field of biofeedback. The Academy of Parapsychology and Medicine is at 314A 2nd Street, Los Angeles, California 94122, and their number is (415) 941-0444.

Robert Mattson led me to two other people, Jack Schwartz of Salem, Oregon, and Dr. Fred Andrews of Redlands, California. Dr. Andrews is actually a foot doctor, but he also teaches astral travel. It seems, however that Jack Schwartz is even more accomplished in this field, and when I spoke with him on the phone recently, he had just got back from ten days of testing by Elmer Green at the Menninger Foundation.

UNIVERSITY OF WASHINGTON: The following letter was received from Dr. Richard Alan Miller of the University of Washington:

Dear Sir,

Thank you for sending your newsletter. I found it most interesting.

You may not be aware of it, but I was just recently appointed Director of Research, Department of Paraphysics/Parapsychology, Experimental College, University of Washington in Seattle.

I currently have a staff of 36; 12 have Ph.D.'s, ranging in fields from physics to social psychology. There are five major programs of study going on, one of which was given in a paper in New York last month. That paper indicated that through the work in Kirlian photography, the so-called "aura" is in part a form of gas emission at discrete points on the surface of the skin. The Kirlian photography technique is a form of corona discharge which ionizes the gases to make them visible to the unaided eye (by the use of the film, presumably.)

I will be giving a paper in Prague in June on the subject of geometrical pattern and how they relate to the electro dynamic theory of life as postulated by Northrup and Burr.

The purpose of this lengthy introduction is simply to inform you of our presence here in Seattle and to offer our help in any way we can to present the "para-sciences" in a more technical and rigorous way. Only through this path will real grants and meaningful research develop.

Sincerely yours,

Richard Alan Miller, physicist.

A PERSONAL PROBLEM AND HEALING: Some time ago, I told of a personal problem with my gums. I asked readers for help, and also wrote to Reverend Willard Fuller, who had apparently healed many dental problems for others in the past. One reader provided me with "Ipsab", which is something that Edgar Cayce's readings recommended, and I used that for a time.

Reverend Fuller wrote that he would pray for my healing, but that in addition, "Twice a day I should brush my gums vigorously with a toothbrush in a certain manner, and while I did this, I should silently thank God for healing them. In seven days—cured!"

I tried to follow these instructions, but after three days of that procedure, my gums were so sore, I could not continue. I then wrote to Reverend Fuller about that, and he replied that I should have more faith, and continue, and he would pray for me again.

—*Key to the Riddle Cancer*, former assistant professor of pharmacology at University of California Medical School, San Francisco, and for more than a decade research consultant to the McNaughton Foundation.

What Some Patients Say

So ingrained is the concept of surgery/irradiation as the sole hope for whipping cancer that a Laetrile patient often has trouble convincing family and friends that the drug has saved his life.

But whether others believe it or not, the patient whose life is at stake usually gives credit to the drug and diet and other elements of the treatment, as well as to the doctors.

Among those interviewed by PREVENTION was Mrs. Joan Wilkinson, 55 Cora Court, Walnut Creek, Calif.

In the spring and summer of 1967, a tumor was removed from her lower left leg, just below the thigh. Two-and-a-half months later there was a recurrence, and surgery was performed in Concord Community Hospital on a malignancy necessitating removal of muscle and bone.

When amputation of a leg, hip and perhaps a kidney were advised in a second encounter with cancer, Mrs. Joan Wilkinson, Walnut Creek, Calif., decided to take the advice of her sister to investigate Laetrile. That was four years ago. She is unreserved in her praise of the doctor and the substance which she believes saved her life.

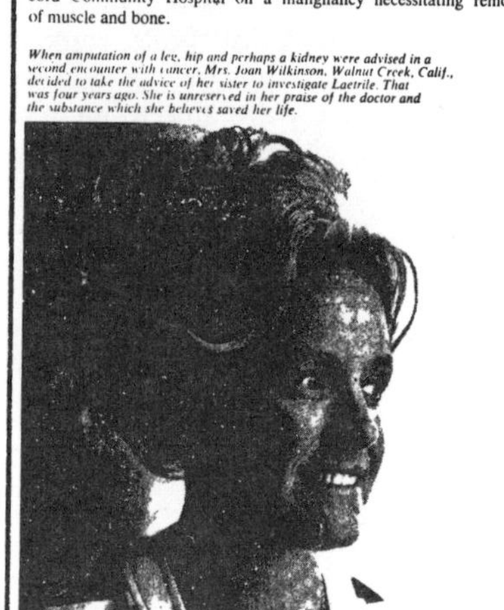

For a time it seemed all was well — but about a year later Mrs. Wilkinson discovered a lump in the groin. She returned to her doctor who ordered a biopsy. Then her physician called to refer her to a surgeon — saying at the same time that it appeared it would be advisable to amputate the leg, hip and perhaps a kidney and the bladder. On the following Friday evening she saw the surgeon who after an examination advised that she enter the hospital Sunday for surgery the following Tuesday.

"The surgeon, who also happened to be head of the Cancer Society, took a book off a shelf and very cruelly told me what parts of the body would have to be removed," she recalls. "He traced the leg and organs with a pencil on an illustration of the human body.

"He also told me that before amputating, he would do an exploratory of the lungs, and if he found cancer there, he would not amputate. As my husband put it, they would sew me up and send me home to die.

"It was a pretty bad time for us. I couldn't sleep. I started praying. Something told me — as if it were God Himself telling me — 'this tragedy cannot happen to me.' It was as if I were another person — 'it is not me this is happening to. I have a family (six children of her own and three step-children) — I have work to do.'

"The next morning my husband called the surgeon's office to ask about having cobalt treatments, perhaps. But the doctor was busy and he didn't reach him. Then my sister, Mrs. Vera Murray (of Orinda, Calif.) called me, pleading with me to go to San Francisco to see Dr. Krebs. She had told me about him and Laetrile the year before, but I had forgotten about it.

"I said to myself, 'I can't lose anything — I'll go.' That was Saturday, November 16, 1968. I'll never forget that day! The stitches from the biopsy were still in the leg. Dr. Krebs gave me an injection of Laetrile — and the tumor reacted, it got very large — from walnut size to the size of a small lemon — and there was bleeding four or five days. I went back on Monday, Wednesday and Friday each week for five weeks to get injections, and the tumor then started getting smaller. Five weeks later I could no longer feel it.

"An x-ray was taken the first Monday, and regularly after that to watch the progress. Injections were continued for six months — 10 cc's three times a week — and of course the diet. No dairy products, nothing made with white flour — no eggs — but white fish, chicken, turkey.

"And I felt wonderful! In fact, in August, 1969, the doctor told me I needed no more injections, my x-rays were clear, showing that the tumor had shrunk, was apparently encased in scar tissue, and was not active. But I worry. I didn't want it to start up again, so I determined to continue with the injections, and still get one a week, and also take it orally, along with some B_{15} morning and evening.

"When the surgeon learned I would not be having surgery, he told me I had from six to eight weeks to live — no more than three months. He pleaded with me to have the operation. I asked him how many operations of this kind he had done — I've forgotten how many he said — but I remember he said two were living and doing very well. He told my husband cobalt wouldn't help in my type of cancer."

The 39-year-old Mrs. Wilkinson, twice a grandmother, says she wishes "Laetrile could be available to all people with cancer. It is too bad it is not easily available to physicians who could use it, check patients' reactions, and report results so that more knowledge of treatment could be obtained and disseminated. If doctors were allowed to work with it freely, learning by clinical experience how much medication should be administered in various types of cases, and when the treatment could be safely terminated, what a blessing this would be.

"I believe each person has the right to do with his body as he wishes. If a person wants Laetrile, he should have the chance to get it. It does offer hope, and that is very important in cancer. When

Continued →

I then attempted the procedure again, and this time, there was no pain; I was able to continue, and have continued to exercise my gums in that way to the present, with a seeming improvement in their health.

When I asked my dentist at the clinic about gum conditions later, he said they were not too bad. He took an X-ray, which he said indicated that there had been no further deterioration, nor any dramatic recovery; that is, the gums had not grown back to where they should be. Later, a more careful study will be done.

CANCER AND WAR: Most of us realize the Vietnam war was a swindle. Few are aware that there can also be swindles going on in the cancer research field. Perhaps what looks like a swindle is, in both cases, only extreme pig-headedness. The suffering is there nevertheless.

For some time, I have reported on the use of Laetrile as a cancer therapy, and now I have further documentation. Recently, Rose Jones of this area went to the hospital in Mexico after she was told that her case was incurable.

After the start of her treatment there, she wrote a letter to her doctor, and on the phone today she gave me permission to publish that letter.

Cancer: A letter from Rose Jones (written to her doctor from Mexico)

Dear Doctor,

I am going to need your help when I get back to follow up the treatment. I am getting the benefit of Dr. Livingston's BCG and the Koch treatment of the world-renowned cancer specialist all given to me by Dr. S. . . . He would like to have a doctor in the Philadelphia area to follow up with the treatment, most of which is accepted in the United States and paid for by my insurance.

I'm staying at this beautiful resort by the ocean for $20 a day, including food and meals on a strict diet of health foods. If you can get me a doctor, I will get and administer the Laetrile myself so he takes no risk, and he could work with Dr. S. on the rest of my program and I will have to come back in six months.

Most patients are allowed one friend to come with them for $10 a day extra, meals included. After the first few days when I began to feel better I could catch a bus for 15¢ which rans past the hospital to Rosarita Beach south or north of Tijuana. If I care to rent a car or take a bus, I could go 50 miles south to Ensanada. The shops are beautiful and just before Christmas one could have such fun.

I have made wonderful friends from all over the country. We all feel that God has brought us here and we're the chosen few of all the cancer victims from the U.S.A. Only six were here last week. Dr. Livingston said that my husband should be proud of me to come here alone and not lean so heavily on him and have courage, trust, and faith in my doctor.

I am sitting in a motel on the border. Dr. S. droped me off last night here. He had to go to Arigonia so I must rush so as not to miss my plane to Los Angeles. When Dr. S. took us to San Diego Monday, he gave his daughter $20 and told her to treat the three of us to dinner and get us a room in a motel to rest until he was free to take us back. You can pass this on to your writer friend; tell him to try to get time on a small radio station to pass on the good word about Laetrile.

Must run, love always,

Rosa

The doctor to whom the above letter was written tells me that Rose's case was far advanced. The doctor did not have her notes at the time of reading me the letter (which I recorded by phone), but she remembered that Rose had had a breast removed, and later, she had lymph nodes removed from the "axcillae" (whatever that is) and a great deal of radiation.

When she returns, she will need to find a doctor to work with who will be open to the therapy she is receiving. I checked up on the identity of Dr. Livingston and learned that she is a very well known doctor who has done research with BCG for some time. A series of articles on this have appeared in the fall 1972 issues of the **National Enquirer** newspaper.

The Koch remedy and BCG are, as far as I know, both methods of increasing a person's immunity to the disorder while Laetrile attacks the cancer cells themselves. More information on the Koch remedy can be had from Borderland Sciences Research Foundation, P.O. Box 548, Vista, California, U.S. 92083. It's 65 pages, price unmarked. $2 should cover it. I will know more after Rose gets back.

LAETRILE AND HEW

(National Cancer Institute)

I have also received a copy of a detailed letter from Dr. Dean Burk, head of the cytochemistry section of the

one is told by a doctor that it's cancer, you almost automatically put yourself on the death list, and your friends do too. I have had friends look at me, their eyes told me they didn't expect to see me again."

With the use of the Laetrile and the diet, Mrs. Wilkinson said she gained appetite at once.

A Prostate Case

Another patient is Joe Botelho, 575 Brookside Drive, San Pablo, Calif., who in the fall of 1968 underwent surgery (transurethral subresection) but not removal of what the doctor told him was a prostate tumor.

"I didn't let them take it out," he told PREVENTION, "because I figured that would only spread it. The doctor told me I wouldn't last too long. He wanted to give me cobalt and I wouldn't agree to that either. At a health food store we heard about a doctor in San Francisco who used Laetrile. I went to see him, was told that the prostate was the size of a bar of soap. I got one injection every four days for several months, from January, 1969 for a year."

The 65-year-old San Pablo man told about the strict diet — no dairy products, lots of carrot juice for several months — says his hair is getting dark again, though he's not sure what's causing it, whether it's the Laetrile, B_{15}, or organic molasses, included in the special diet.

Joe Botelho spends a fair amount of time on his tractor doing truck gardening or working around his place in San Pablo, Calif. After a transurethral subresection for what was diagnosed as prostate tumor, he refused cobalt treatment and started on Laetrile in January, 1969. He remains in good health, watches his diet — no dairy products, plenty of carrot juice and organic molasses.

Life-Saver, She Says

Another who has been "on Laetrile" since January, 1967 — a woman who has lost her parents and a brother to cancer — is Mrs. Earl Clark, 5275 Olive Drive, Concord, Calif. She, too, is convinced she owes her life to the controversial substance.

When told in October, 1966, that she had a tumor in the cervix, she decided against a hysterectomy. She was considering going to Tijuana to see Dr. Ernesto Contreras "in a couple of months," when fate intervened: Her father-in-law, the late Frank Clark of Concord, had been given three weeks to live — he was suffering from rectal cancer — and Mrs. Clark accompanied him to Mexico. (Mr. Clark lived 18 months, died of a heart attack). Mrs. Clark was examined while she was there. A urine test "was very high," and she started taking Laetrile capsules, continuing 12 months, when two urine tests three months apart "came back negative." The tumor condition continued to develop, however, and in July, 1968, Dr. Contreras and a colleague performed surgery, removing "a quart of tumors," which biopsy revealed to be non-malignant.

When he did the post-operative examination a month later, her gynecologist in Concord, the late Dr. Lee Lorenzen, described the Mexican doctor's work as "excellent surgery," Mrs. Clark says.

For a year she continued feeling well, then started losing weight. Alarmed over the loss of 18 pounds in two weeks, she returned to Tijuana where she stayed three weeks, receiving "massive doses of Laetrile." She went back on the rigid non-dairy products-fruit diet, takes her prunes every morning as advised by the late Dr. Krebs, Sr. and the tests show a continuing reduction of cancer activity, she told PREVENTION. Mrs. Clark also has a tumor on the thyroid, lives with the knowedge it may some day have to be removed, "but I'm not worried about it because with the Laetrile treatment I know it won't be malignant." She still intends to continue — "Dr. Krebs told me to never go off it completely."

Anonymity Desired

PREVENTION talked with two other Laetrile patients, neither of whom wish to be identified because of personal or family reasons.

One is a woman in her early seventies who has been getting Laetrile since 1932 when the late Dr. Krebs was doing his early work with the substance. She had a lump in the breast, refused surgery. She took injections twice a week for three months, then twice a month for six months, and less frequently over a period of three years. "Then there was no more lump, except the cartilage around it." Another lump appeared in a different location, the size of a hazelnut, so she started in again with Laetrile. It was reduced to the size of a pea, and a test eight months ago revealed it as "positive," while three months ago the urine test came back "negative." She remains on a limited treatment schedule — one injection monthly, three small tablets in the morning.

Hodgkins

The other patient is 24 years old, daughter of a prominent executive whose business brings him into contact with top doctors in the country. He and his wife are "sold" on conventional therapy. Their daughter developed Hodgkins disease, diagnosed in June, 1966, shortly before she was to have a baby. She learned about Laetrile and vitamins through a "hippy-type" friend.

Continued →

National Cancer Institute (HEW), Burk charges that the FDA has been refusing to allow the testing of Laetrile here on the grounds that "it might be poisonous," whereas his tests show that Laetrile is the least poisonous of all cancer therapies now known. He has charged the FDA with dishonesty. See excerpts from Burk letter later.

I then spoke with Dr. Sidney Wolfe, who is the head man of the medical section of Ralph Nader's consumer group, and sent him a copy of that letter. Dr. Wolfe told me that the FDA is now "reconsidering" the use of Laetrile (which is already being used successfully in 22 countries including England, Germany, and Mexico). He sent me a copy of a letter received from Earl L. Meyers of the FDA, in which he tells of Dr. Contreras (Mexico) and Dr. Nieper (of Germany) working with Laetrile, and of 21 cases whose records Dr. Contreras has submitted. The FDA offices are working on an investigation of these records. I have some doubts as to their eagerness, however.

CANCER: Excerpts of Burk letter:

"NEW FOODS CONTAINING LAETRILE:

Recently, starting in Europe, efforts are underway to produce foods rich in Laetrile, where (a) it will not be a drug, and (b) can be consumed in sufficiently small food quantities, yet provide current daily recommended Laetrile dosages of the order of several grams per day, at approximately one-tenth of the cost per unit of Laetrile now otherwise prevailing. Thus flours may be made from bitter almonds (extensively available in Europe—but outlawed here by FDA), or apricot kernels (extensively available in this country), or other kernels of the prunus family. A variety of foods are under consideration, including bread, pastry products, even milk shakes. The bitter almonds themselves have long been available in European grocery stores, just as the apricot kernels have been here.

Other Laetrile possibilities: There are about two million persons in the United States with the sickle cell trait, and some 50,000 with actual sickle cell anemia, which very recently has been found to be ameliorated—at first by urea—and now much better by an impurity in urea, namely, cyanate. Tablets or capsules of cyanate may be ingested to overcome the anemic hemolytic crisis. Cyanate can also be produced by hydrolysis of Laetrile, and conceivably it might be more sensible to obtain cyanate from ingested Laetrile—a working hypothesis yet to be demonstrated as efficacious, be it clearly noted, but which, if demonstrable, could be demonstrated with far less scientific effort than has been involved with Laetrile and cancer. This approach has been suggested to me by one of the most brilliant biochemists I know of in this country, who writes me further regarding this hypothesis: "A relatively few simple clinical applications of amygdalin (Laetrile) in pure form or Laetrile in good rations should give us the answer. If the answer is confirmatory, I wonder if the FDA will attempt to perpetuate a discriminative genocide against the black population of this country who have sickle cell anemia." He adds that "Populations in Africa who have diets rich in vitamin B-17 (Laetrile) do not have sickle cell anemia, no matter how loaded they are with sickle cells." He further points out that there appears to be a remarkable difference in cancer incidence among Nigerian natives who do and do not have large amounts of vitamin B-17 in their diets, greatly in favor of those who do, with respect to absence of cancer, according to reports by Dr. O. L. Oke, University of Ife, Ibadan, Nigeria.

Sincerely yours,
Dean Burk, Head,
Cytochemistry Section,
NCI"

DISCLAIMER: The above views of the oversigned research scientists may differ materially from those of administrators within HEW, and no such latter official support or endorsement is intended or should be inferred.

HIPPOCRATES HEALTH INSTITUTE

I was recently accompanied on a trip to the Hippocrates Health Institute in Boston by a doctor who spent five days at the Hippocrates Health Institute. Her purpose was to find out if the live food natural diet really was helpful in the recovery of "terminal cases." Results of the study seem to indicate that some people have indeed been restored to good health who had been entirely given up by their physicians. However, it was often so that the Institute did not do sufficient follow-up on the people who came, so that there were only a small number of cases where the fact of a cure could be established. At the same time, it was found that most people found the new diet so difficult to stick with that they did not continue it sufficiently to give it a fair trial.

In short, the evidence seems to indi-

"I first knew something was wrong when the bottoms of my feet and legs started itching," she said. "The doctor said it was neurodermatitis. He said I was emotionally upset and my parents sent

Convinced that she owes her life to Laetrile is Mrs Earl Clark, Concord, Calif., who was told in 1966 that she had a tumor of the cervix. She took her father-in-law to Tijuana for his own treatment, then decided to start the treatment herself. A year-and-a-half later she submitted to surgery in Mexico. Dr. Ernesto Contreras and a colleague removed "a quart of tumors." She remains on Laetrile treatment and a strict diet.

me to a psychiatrist who said there was nothing wrong with me.

"I ate the usual kinds of food including lots of white-flour pastries, sweets, the typical American diet. Then about two months before the baby was to come, a lump appeared on my neck. I thought it was a glandular thing which pregnant women sometimes get. My gynecologist sent me to a dermatologist who sent me to a surgeon who told me it was Hodgkins disease. X-rays showed a whole chestful of tumors; a biopsy confirmed malignancy.

"I had been planning to have a natural childbirth, and you can imagine how I felt when seven doctors came into the room after the biopsy result was known, and told me I should have a Caesarian, to save the baby. They said my disease was incurable, but they might prolong my life. Against my will, I agreed. I had a spinal and could hear the medical students (at University of California Medical School, San Francisco) talking as they watched the surgery. I was supposed to be a rarity, because of such a strong body, despite the Hodgkins disease. The baby was very healthy. At the same time, the doctors took biopsies of liver and spleen and it was found the disease had not gone below the diaphragm."

Insistent upon nursing the baby, she refused chemotherapy, but under pressure from doctors and family, agreed to take cobalt treatment, describing that experience with a shudder:

"I went to U.C. Medical School Hospital to get measured for the cobalt treatment. While I was sitting in the wheel chair, I suddenly had a premonition that all was not right. I got up, it wasn't easy either, and walked out, into the street. My frantic mother came after me, pleading for me to return.

"Treatment was to have started the next day. Then followed more persuasion from family and doctors, and I agreed finally to take the cobalt — at Mount Zion's Claire Zellerbach Saroni Memorial Tumor Institute. When I went in, they started measuring me again and when I asked them why, that it already had been done, they told me it had to be done over. What do you think — the first measurements put the lead block several inches away from the right spot, and would have exposed my heart to the cobalt. I'd have died!

"In the next six weeks I received 4,200 rads. Half my hair fell out; I think it was affecting my brain — and I knew things were not the way they should be. The waiting room was like Death Row, really. Everything is in blue — and people sit there waiting to go in for treatment. They tell about their problems — it's sad. They tell about how the doctor has said 'this is the only thing that will save me' — as they lose another part of their body.

"I was getting angry with the whole thing, a childish response perhaps, but I started refusing blood tests. Just tore up the slips as I walked out. Another test that I refused was the lymphangiogram — where they stick needles in your feet — the test takes an entire day — dye is injected — it really hurts. And you're green for about a month.

"As it was, I had burns all over my face. I learned about using aloe-vera and wheat germ oil in combination for radiation burns — and started using it. It helped. I told other patients who also started using it, and I asked doctors why they didn't tell people about it too. They called it baloney, wouldn't have anything to do with it."

The lady became a gadfly and the medics didn't hide their irritation, she says. It was at that time she heard about Laetrile and diet — in April, 1968. She used the chystalline form — eating it instead of diluting and injecting — and started on a diet not unfamiliar to PREVENTION readers. And vitamins.

She credits the combination of Laetrile, vitamins and diet with "helping my body repair itself."

Her parents have not yet accepted the fact she appears to be recovering, and it is her hope that one day — "when I know I am completely well" — she will be able to confront them, say in effect — "look, they said I couldn't live, you remember that, but I'm not dead, I'm alive, and well. It had to be Laetrile, diet, and a spiritual birth."

Alive and Well

The case of an 18-year-old boy who had what was believed incurable cancer of L3-L4 level which spread from a testicle to the lungs — his physician said he was "not confident" that any form of therapy would interrupt the progress of the illness — makes another incredible history. (The young man's father requests anonymity).

From surgery to remove an embryonal cell carcinoma on the right testicle in September, 1966, the cancer spread to the lungs, and chemotherapy — 15 gamma per kilo of Actinomycin-D for five days — was administered. Since the pulmonary nodules had increased in size during this course of therapy, his doctor urged further chemotherapy in the form of a "combination program including methotrexate and an alkylating agent, probably Cytoxon." But the father — described by the physician as "an extremely intelligent but deeply involved individual" — balked, and took the boy to Dr. Contreras in Tijuana.

The records are documented: After massive doses of Laetrile

Continued →

cate that Ann Wigmore's discoveries could lead to a tremendous health breakthrough, providing good nutrition at amazingly low cost and restoring many people to good health. But more research needs to be done to substantiate that, and the diet needs to be introduced in ways that are more acceptable to the average person. I believe that Dr. Wigmore would be open to her staff being enlarged by some open-minded and capable researchers who could help her to bring that about. People who are interested could start by writing for information to the Hippocrates Health Institute, 25 Exeter Street, Boston, Massachusetts 02116.

TRANSMUTATION OF ELEMENTS: New sciences, indeed! Or is it just the revival of alchemy? I recently reported on an experiment in which it was shown that it may be possible to transmute the element, carbon into iron. Bruce Gardner in Boston told me that he has done the experiment successfully, but also admitted that it is conceivable that there are tiny particles of iron in the carbon rods which are used. I would like to suggest that this experiment be repeated under modern laboratory conditions in which it is actually possible to analyze the copper and carbon for traces of iron, and where the purest materials could be used. If it is true that transmutation is as simple as claimed, then the chemistry of our time is seriously lacking.

To obtain detailed descriptions of the experiment, send $1 for a paper called, "Order of the Universe," Volume IV, No. 8 to Tao Publications, 29 Farnsworth Street, Boston, Massachusetts 02210.

BIOLOGICAL TRANSMUTATION is the title of a very well written book by a French scientist named Louis Kervran. I have not yet had time to read the whole book, but what I have seen has convinced me that Kervran is a very knowledgeable scientist who has studied the question of the transmutation of the elements in the most careful manner in line with modern science, and has conclusively shown that the transmutation of one element into another is actually occurring in nature and in our bodies each day. He has demonstrated this to scientists in France, and has been invited to China and Russia (where all his works are being translated) and to the United States by Dow Chemical. (He has not come yet.) The book is written in such a way that it is understandable to anyone with a knowledge of elementary science. The book can be obtained from Tao Publications for $2.75, or from Swan House, Box 638, Binghampton, New York 13902.

PARAPSYCHOLOGY IN COLLEGES: With the rise of parapsychology as an accepted science, and astrology becoming a field for careful scientific investigation, more and more colleges and universities are sponsoring classes and conferences in these subjects.

Some of you might have wanted to attend an interesting symposium on astrology under the auspices of Boston University, held at Boston University and co-sponsored by ARC (Association for Research in Cosmecology, c/o Chas. Jayne, 5 Old Quaker Hill Road, Monroe, New York 10950).

The symposium had panels on: astrology and education, astrology and the humanities, astrology and the sciences, and astrology and psychology. Guests of honor were the Pakistani Consul-General to the U.S., and the president of the Astrological Association of Great Britain. Other speakers were: Arthur Blum, M.S., American Institute of Physics, Miguel Charneco, Ph.D., Psychology Department of the University of Puerto Rico, Zipporah Dobyns, Ph.D., Thomas Glynn, Ph.D., Executive Secretary for the National Council for Geocosmic Research, Arthur M. Young, Director of the Foundation for the Study of Consciousness (and inventor of Bell helicopter rotor), Baldur Ebertin, Ph.D., Academy of Kosmobiology (Eastern Europe), and many others.

ESP CONFERENCE: On the 2nd of March, 1973 there was a conference on ESP held at the Blue Ridge Community College in Virginia, and some of the best researchers were there. They were: Robert Von De Castle, Director of the Sleep and Dream Laboratory at University of Virginia, Douglas Dean, Director of the Psi Communication project at Newark College of Engineering, Dr. George Ritchie, Jr., M.D., Central Virginia Psychiatric Association, Robert Monroe, Director of Metronics Research

through January, 1967, a film of January 31 that year showed reduction in the size of the hilar nodes as well as the nodules in the apex and second anterior interspace.

Two weeks later the x-ray report (by a San Francisco firm of M.D.s) revealed "further improvement. The nodular density in the left apex has decreased in size and is not as evident, and I no longer can see a nodular density in the second space anteriorly on the left, nor along the lateral thoracic wall at the plane of the sixth rib. The left hilar shadow, while still slightly prominent, has an essentially normal configuration and there are no nodular densities now evident in the left hilum. The right hilar shadow is not remarkable on this study and there are no nodular densities in the right lung."

X-ray film by the same firm 10 days later (Feb. 17, 1967) showed "further decrease in the nodule in the left apex," and "no definite abnormalities recognizable in the hilar structures," but "a vague density overlying the left fourth rib anteriorly projecting between the posterior sixth and seventh ribs, too indistinct to determine whether it is a significant lesion or not. The rest of the lung fields appear clear."

And on, April 11 that year the laboratory report came back: ". . . Complete clearing of the hilar adenopathy and pulmonary nodules. No lesion is identifiable in the chest at this time."

The young man has periodic checkups at Stanford University Medical School, and after an absence of one year from his studies, returned to school and is still very much alive.

There are 5,000 case histories similar to these we have cited after investigation. Yet, the panel of cancer experts convened by the FDA when it announced its decision to continue the ban on clinical trials of Laetrile on humans said there was "no evidence to warrant further testing."

The basic issues as we see them are the deprivation of freedom of choice by the patient, and the freedom of choice of their physicians of the methods of treatment available in cancer.

Even worse, people have been arrested in the state of California for helping other people — cancer patients — to get to Tijuana, Mexico, where they could legally receive Laetrile treatment. In Pennsylvania a health club meeting was raided when a film advocating Laetrile treatment was shown.

Such actions never stand up in court, of course. They are clear violations of personal liberties guaranteed by the U.S. Constitution. Yet they remain effective harassment calculated to intimidate people trying, rightly or wrongly, to help other people who will surely die without their help.

Why?

Is this country genuinely interested in preventing and curing cancer, or is it really aiming only at a more limited objective? Is it really trying only to find a somewhat more successful treatment that will keep patients alive longer before they succumb, but will still keep all the surgeons and radiologists and chemotherapists driving their new Cadillacs to the best country clubs?

There is no doubt the entire medical profession would welcome better cancer therapies than are now available. Yet when something appears that holds out the hope of *preventing* cancer even more effectively than it treats it, suddenly the same people who have been treating cancer with violent poisons for 20 years become alarmed about possible toxicity; they become blind to well-authenticated case histories; after extracting billions from Congress and the public for research, they determine that experimentation with the informed consent of the patients would be unethical; and they show an utter disregard even for the Constitution of the United States in their efforts to make it impossible for anyone even to discuss the new treatment.

Why?

The doctors Krebs believe — without conclusive evidence — that primitive people eat the seeds of such fruits as apricots, peaches and apples, and that those seeds in effect immunize against cancer. We can't say whether they're right or wrong. But if we had a friend who wanted to try, we wouldn't discourage him. Our strongest feeling is that you have a *right* to eat apricot kernels if you believe they are good for you.

When the government tries to say you have no such right, something is wrong. □

DISTRIBUTED BY:
The Committee for Freedom of Choice
In Cancer Therapy
146 Main St. Suite 408
Los Altos, California
94022

OBJECTIVES OF THE COMMITTEE:

1. The Committee does not claim competence to judge the relative merits of various cancer therapies.
2. The Committee *does* take a strong position, however, on behalf of freedom of choice for both physicians and patients to use whatever therapy they mutually accept.
3. History has shown that governments often have institutionalized and perpetuated error in scientific questions. Therefore, the Committee opposes any attempt by the FDA or any other agency of government to pass judgment on any cancer theory or therapy.
4. All currently used cancer therapies–surgery, drugs, and radiation admittedly are dangerous and harmful to the patient in varying degrees. Therefore, the Committee does not support the FDA's ruling against the use of Laetrile (a non-toxic vitamin food supplement) on the grounds that it *may* be harmful.
5. The quality of medical research and practice declines in direct proportion to the degree to which it is influenced by politics. Therefore, the Committee believes that the government should get off the back of private research and out of the field of medical practice.

The Committee pledges its moral and financial support to physicians and patients in their legal battles to achieve these objectives.

Institute at Afton, Virginia, and author of **Journeys out of the Body**. Also attending was David Hoy, psychic, and author Harold Sherman, and others.

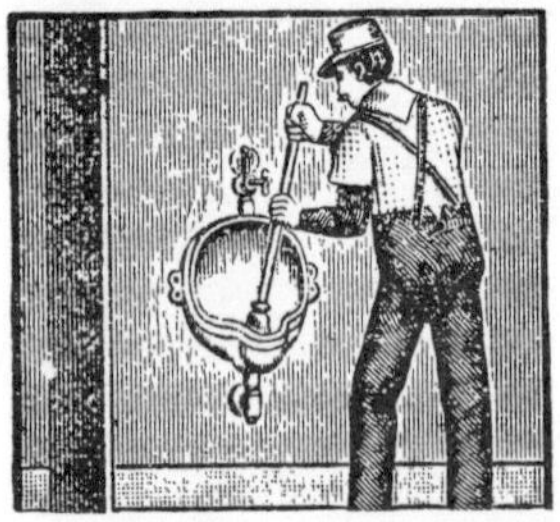

NATURAL VITAMIN C BARGAIN As many of you know, a vitamin C supplement is virtually necessary in any urban area where the high concentrations of pollution quickly deplete the body's normal store of vitamin C. Ascorbic acid is a chemically derived substance duplicating the major constituent of the natural vitamin C complex, and is far cheaper than an all natural vitamin C supplement. However, it is best that one-tenth of your vitamin C supplement be natural to supply the rest of the complex.

Rose hips contain the largest percentage of natural vitamin C we know of. Rose hips are available from Walnut Acres, which has had a very high reputation of long standing in the organic food business. They list rose hips powder on page 5 of their current catalogue under teas for $1.28 per lb. We have been using it and have felt it to be effective. For their catalog send 2 stamps to Walnut Acres, Penn's Creek, Pennsylvania 17862.

The **New Spiritual Community Guide for North America**, 1975-76 is a book containing over 3,000 city-by-city listings of yoga and meditation centers, ashrams, natural food stores and restaurants, bookstores, etc. It also has almost 100 descriptions of major spiritual groups in the United States. It can be ordered for $3.50 from Spiritual Community Publications, Box 1080, San Rafael, California 94902.

ATOMIC POWER FROM WHOM? Senator Mike Gravel of Alaska is one of the few people in government who is really thinking about the power question realistically. He publishes a newsletter which is sent out on a free or (preferably) donation basis.

Senator Gravel favors various types of solar power over the present trend to atomic reactors which is being pushed by big industry and the AEC, quite regardless of the obvious risk of accident and the continual danger from harmful radiations. He says:

"Few members of the present Congress have ever had to think about nuclear energy. It was largely a different set of individuals who passed the Atomic Energy Act in 1954 directing the atomic energy commission to promote civilian nuclear power. For 20 years, there has been no debate of vote in Congress on that fundamental decision. In effect, Americans under the age of 40 have never given their own consent to our hazardous experiment with civilian nuclear fission."

As evidence for the belief that nuclear reactors are not safe, Senator Gravel points to the fact that the insurance industry which is experienced in the study of this question steadfastly refuses to sell more than a limited amount of insurance to the atomic power industry, and then charges a very high premium.

At the same time, the power companies are not paying for the insurance, but this is paid for by the U.S. government under the Price-Anderson Act, which Senator Gravel wants repealed. Seems that if the companies had to pay the insurance, they may not build more plants, even though they claim accidents won't happen. As Gravel says:

"Meanwhile, the insurance industry has voted no-confidence in 'safe' nuclear power by refusing to insure our homes, business, and autos against damage from nuclear power plants. Policies contain a special 'exclusion clause'."

"As for providing public liability insurance to nuclear utilities, even a pool of insurance companies today cannot be persuaded to offer more than $110 million per accident. In exchange for this $110 million in public liability coverage, a utility pays an annual premium between $250,000 and $450,000 on each nuclear plant, according to AEC Commissioner Doub."

SOLAR ENERGY: Dr. William Hieronymus of the Department of Civil Engineering of the University of Massachusetts in Amherst has developed a method for producing power from the Gulf Stream in such vast quantities that it can replace the entire electric generating capacity of this country six times over, without cooling the Gulf Stream more than one-tenth of one degree.

It is known that the sun heats the waters of the Gulf of Mexico, and causes a stream of warm water to flow across the Atlantic. This water is 30° warmer at the surface than it is down below.

They have designed a power plant that will operate below the surface of the Gulf Stream in what might look like a large stationary submarine, with an entrance tower rising 20′ above the surface of the sea. On top of the plant, heat exchangers will carry liquid gas, which will be heated by the warm surface water. These will be located 20′ below the surface. Well beneath the power plant, there will be the condenser units in the deep, cold water below. The higher temperature will cause the gas to expand and operate the generators inside.

Each power plant will have 16 generating units, producing 25,000 watts of free electricity each; a total of 400 megawatts per power plant. That is possible because within the Gulf Stream, 40,000,000 cubic meters (a cubic meter is about 40 cubic feet) of warm water go by each second! It is calculated by Dr. Hieronymus that it would be possible to locate as many as 5,000 such power plants in the region of the Gulf Stream, and the electricity could be piped to the shore by cable.

THE COST: Careful cost calculations have also been done, and it has been shown that a pilot plant generating 400 megawatts continuously without fuel could be built for about $400,000. Later plants could be built at about half that cost. At a price of four cents per kilowatt hour (which is about what individuals pay for electricity), such a plant would pay for itself in three years' time. From that time on, the power would be completely free (except for the costs of maintenance and distribution, which would probably be less than the cost of maintaining current equipment).

At present, the U.S. government is already planning to spend $750,000,000 on research alone to produce nuclear breeder reactors, which are known to be extremely hazardous with radioactive wastes. Some have suggested that that be dropped in favor of the Gulf Stream project.

ENERGY AND LIGHT: The Divine Light Mission. What is the Guru Maharaj Ji? Is God a fifteen-year-old "Perfect Master"?

Several people have asked me to investigate the Divine Light Mission and find out about the "Knowledge" that they give, as well as their hopes of bringing about the miracles that will change the world.

Accordingly, I have spent some hours listening to "Satsang" (spiritual talks) from various members, and have also spoken with one enlightened individual who did not consider himself to be a "humble disciple of the Lord." I have attempted also to receive the "Knowledge," but was, for the present, refused.

In recent times, the "Knowledge" was easy to obtain, as it was given freely to whomever asked for it. But since it was discovered that only one out of five who received it continued faithfully in the meditations, and followed the teachings of the organization (which is believed by many of the followers to be perfect), they have decided to limit severely the giving of their Knowledge.

Hence, the brother of the Guru has told the "mahatmas" that they may only pass on their "Knowledge" to 15 people at a time. As a result, they have many people following a mahatma around from city to city and listening to the Satsang until such time as they have heard it about 30 times and are ready to receive the Knowledge. I did not feel able to drop my work to do that now, and it appears that exceptions are rarely made.

After hours of waiting, I was allowed to speak to the mahatma to present my case, but I had the impression that he did not understand enough English to follow what I was saying. He also interrupted from time to time and continued to insist that to receive Knowledge, one must first hear 30 Satsangs and read their publications more than I have done.

Nevertheless, some of what I saw in their communal ashram life was impressive, even though the organization was very much of a "top-down" kind of hierarchy. Since the followers all appear to accept the view that the 15-year-old "Perfect Master" is actually God himself, they are perfectly ready to accept whatever word comes down to them and do whatever they are asked in a loving and grateful manner. Having myself spent 15 years in a similarly organized community, I can see the beauty of that kind of life, as well as some of the possible fallacies, which may develop in the event that the Guru turns out not to be as perfect as his followers believe him to be.

One of the most moving experiences among them was to meet and listen to Rennie Davis (of Chicago Seven fame), whom I had known before the trial in the late 1960's. Rennie has experienced a conversion to the Guru, and is travelling around the country on a speaking tour and mission with the theme "Who is Guru Muharaj Ji?" He is asking us all to find out for ourselves who this Guru is by experiencing his Knowledge, which is also described as an "opening of the third eye." Of course, Rennie did not have to follow anyone around for weeks to receive the Knowledge, but instead was given a free ticket to India. I don't know if he would ever have investigated the matter if it had been as difficult to do then as it is now.

Nevertheless, I was impressed by Rennie's openness and honesty. I have no doubt that he has experienced many miracles since he has aligned himself with the work of the mission. But it has also been my own experience that anyone who drops all else and puts himself fully into the work of bringing God's Kingdom, the Aquarian Age, or what have you, experiences one miracle after another from that time on.

So we are still left with the question, "Who is Guru Maharaj Ji?" Well, who does the kid himself say he is? I am told by one who received the Knowledge that he is a saint, but all that he says is "I'm a groovy kid that's got something you will really like!"

There is evidence that many who receive the Knowledge that he and his mahatmas are able to give an uplifting experience, leading to deep spiritual insights, and the dedicated followers seem convinced that they are going to be used to change the world.

A SILVA MIND CONTROL SCHOOL: It came to me in conversation with Jon that it would be a fine thing if some of the instructors and graduates of the Silva program would start their own private schools in which young people, even children, might begin by learning the mind control method as the first part of their school training. Then they might continue using these methods in order to learn all the school subjects, using the heightened awareness and better concentration which the Silva method would give them. They should then be able to learn all they need to know to pass college entrance exams at a very early age, and perhaps some of them could then whiz through college and start their own schools, etc. etc.

In a talk that José Silva gave during the advance course that he taught here, he suggested that one of the troubles with modern education we see in schools today is that when the student is old enough to be physically mature and raise children of his own, he is totally unprepared to do so from the mental or confidence or stability point of view.

As a result, the child grows up physically but not mentally until he is very frustrated at not being able to DO anything that is really meaningful.

With the Silva method of learning to use all of one's natural mental levels in creative ways, the child (so José thought) might be able to be a man or a woman at about the same time as they arrive at physical maturity, or even earlier; that is apparently the way it was in biblical times, when a boy of thirteen was definitely declared to be a man. The custom (called Bar Mitzvah) still prevails today among Jewish people, but the boys are not ready for it. Perhaps when you read this, José, you might consider starting such schools. (And that could apply to all you other mind control folks also.)

RISKS OF PSYCHIC RESEARCH: Last Saturday, I spent most of the day driving between Philadelphia and Harrisburg with an engineer in order to check on a claimed invention of another "fuelless engine" by a man named Russell

Brown, which had received rather wide TV and newpaper coverage. We found Mr. Brown to be an honest person, but were able to determine that he did not actually have an invention that seemed likely to overcome the need for fuel as a motive power source. He seemed to lack the scientific background to accomplish the ideals which he has set for himself.

During the trip, we spoke about various psychic experiences and were reminded of the danger, which is very real, in the pursuit of knowledge of the spirit world, especially when such knowledge is sought for reasons other than a pure and sincere love for the creation and its purpose.

A recent communication from the community of the Society of Brothers in Connecticut (Evergreen Community, Westside Road, Norfolk, Connecticut 06058) brought this out in no uncertain terms. Members of that group appear to speak from long experience in this area, and reinforce our own conclusions that the spirit world really exists. It is not a thing to be taken lightly or played with out of curiosity or egotistical purposes.

CANCER: FDA vs. APRIKERN & B-17 A few days ago, it was announced that the FDA had found that two products of General Laboratories (California), which are sold in health food stores, may cause poisoning if taken in large quantity. They claimed that Aprikern, or B-17, contained enough cyanide (which is a natural constituent of apricot kernels) to kill a child if 5 were taken, or kill an adult if 20 were swallowed.

Since I had met the director of General Research Laboratories (6925 Hayvenhurst Avenue, Van Nuys, California 91406), and eaten a quantity of Aprikern, and mentioned it earlier as a cheaper form of Laetrile, I was concerned to find out about these accusations.

I telephoned Dr. Dean Burk of the National Cancer Institute in Bethesda, Maryland (301-496-3339). He is head cytologist of that organization, and has been with them over 35 years. We talked for some time, and he told me that Aprikern is not poisonous, and that no one is known to have been poisoned by it. As far as he was aware, a biochemist in Arizona calculated the amount of cyanide in the product and theorized that it was possibly lethal, disregarding the fact that it is so locked together with another molecule that it cannot be a cause of harm. Burk said that cyanide is also found in meat as a natural constituent, and does no harm.

He further declared that he has been conducting an animal study with Aprikern by feeding it to mice, so that 50% of their diet is that product. They have been living on it for ten months and are perfectly healthy. He then offered to mail me additional information on other recent and promising research.

Dr. Burk also agreed with my opinion that in cases of cancer, any treatment that dramatically improves the general health of the patient is likely to make it possible for the body to get rid of the cancer.

YOUTH LIBERATION There is a young peoples' group called Youth Liberation that is now printing a significant magazine called **FPS** ($6/yr. if under 18, otherwise, $10/yr.) They have also published a rare book about the problems of that vast minority written by young people themselves. The book, **YOUTH LIBERATION**, ($1.70 incl. postage) and the magazine

Weekly News Report

CONGRESSMAN

JOHN G. SCHMITZ

For release August 9. 1972 | 72-32 | NOT PRINTED AT GOVERNMENT EXPENSE

LAETRILE: THE CANCER FIGHTER WASHINGTON WILL NOT TEST

"In spite of the aforementioned Food and Drug Administration (FDA) prohibition of Laetrile in interstate commerce, there are well over 1000 cancer-afflicted persons in this country using Laetrile for cancer treatment and amelioration, and a goodly number of non-cancer persons using it merely with prevention of development of cancer in view, and these various persons include M.D. physicians as well as laity. I have had considerable personal experience in this regard, for in the past year alone at least 750 persons, including more than 50 physicians, have contacted me for information on the use and availability of Laetrile, and I know of others with approximately the same quantitative extent of similar experience. In over 20 countries of the world, well over 5000 cancer patients have been treated with Laetrile, with, significantly, no demonstrable noteworthy clinical contraindication of its use either alone or in conjunction with virtually any other anti-cancer agents, chemotherapeutic, radiological, or surgical. Laetrile at physician-prescribed dosages is nontoxic by a factor of 100-1000 times when compared to essentially all anticancer drugs now used with FDA approval...

"Although the foregoing Laetrile utilization in this country is proceeding, as indicated, in spite of FDA prohibition, it is even more so because of unwarranted FDA procedures, and lack of FDA scientific and medical justification for its stand, extending to probably unconstitutionality, concerning which many thousands of cancer-afflicted persons and their relatives and physicians are rapidly becoming aware." -- from a letter to Congressman Louis Frey, Jr. from Dr. Dean Burk, head of the Cytochemistry Section, National Cancer Institute, May 30, 1972

In the case of other drugs and medicines removed from the market, as discussed in my newsletter of July 26, 1972, at least some medical tests were cited to show "ineffectiveness." But in the case of the anti-cancer agent Laetrile, the Food and Drug Administration refuses even to make clinical tests! They have ruled Laetrile out because it does not fall within an arbitrary definition that any anti-cancer drug must be toxic -- that is, to some extent poisonous to healthy cells as well as cancer cells. Incredible as it may seem, some medical bureaucrats have simply decided that no non-toxic substance can ever help cancer patients, and therefore all such substances are to be automatically excluded from the market and no medical evidence in their favor is even to be considered.

If there had been an equivalent of the Food and Drug Administration a century or more ago, it might well have banned, at an enormous cost in human suffering and mortality, both antiseptics and anesthetics, which were strongly opposed by influential elements in the medical community of that day.

This is not a matter of "quackery," panaceas or "secret remedies." The head of the Cytochemistry Section of the National Cancer Institute is hardly a quack. Far from objecting to testing, or blocking it, this is what Laetrile advocates are asking for, while the supposedly "scientific" medical establishment is refusing to make the experiments which are the very basis of the scientific method.

I have introduced H.R. 12092 which would override the arbitrary FDA dictate against testing non-toxic anti-cancer agents and require that this testing be undertaken. This would seem the least that should be done in light of the evidence that Dr. Burk and others have assembled showing that Laetrile can prolong and in some cases save the lives of cancer victims, and the generally agreed upon fact that it does no harm to the patient, while conventional toxic anti-cancer drugs often have serious side effects precisely because they are toxic.

Millions of dollars are to be spent under the authority of recently passed legislation to search for better treatment or cures for cancer. It would seem that a little of all that money ought to be allotted to find out if Laetrile might be what they are looking for.

California 35th District

1208 Longworth House Office Building
Washington D C 20515

Permission is hereby granted to quote all or any part of this newsletter, with credit

can be had from: Youth Liberation, 2007 Washtenaw Avenue, Ann Arbor, Michigan 48104.

CHANGING THE WORLD

Many of the things which I have researched and written about over the years might help to open peoples' minds so that we can ease more gently into the times to come. Re-awakening our natural psychic abilities may improve communication between peoples and may also help us to experience the physical reality of "mysterious" new sciences.

But the most important ingredient that will make all these things happen is not found in science alone; not even in communal living. The greatest need is for individuals who wish to change the world to become effective. This requires a healthy change of attitude in which a person becomes less concerned with his own problems and more concerned with the need of all mankind. The road to effectiveness might be briefly described as follows:

1. Large changes are often due to small groups of dedicated persons. Try to be one.

2. See the goal. Try often to visualize the new world as you would like it to be.

3. Ask "What is the first little step I should take to head in that direction?"

4. Realize that you cannot envision the second step accurately until you have already taken the first step. Acting upon the insights you have results in further insights.

5. "Earning a living " first is not primary. Experience demonstrates that putting the New Day first is actually a direct road to solving your own problems: "Seek ye first the rulership of love (Kingdom of God) and His righteousness and all these things shall be added unto you." Help often comes from the most unexpected sources.

6. People who are effective are generally those who believe in what they are doing because they are doing what they believe in. If present activities have little meaning for you, do something different.

7. Doing these things may seem difficult until you start to do them. Once done, they seem ever so easy; and when you become effective, you find that it is great fun besides.

PSYCHIC-URI GELLER AND UFO: The June 1973 issue of **Psychic** ($1 Box 26289, Custom House, P.O., San Francisco, California 94126) carried an interview with Uri Geller, and an article about him by Alan Vaughn. Both of these pieces are well done and important to read, but one part was disturbing. In Alan's article, he quoted Dr. Puharich as he told about an incident in which Uri led him to a place where they found a flying saucer in a field and Uri told Dr. Puharich to wait for him as he went and entered into the machine.

(This took place in Israel where Dr. Puharich was studying Uri Geller, to find the "world's best research psychic" that he could find. Dr. Puharich is a medical doctor and a scientist who has done a great deal of scientific study in the psychic field and has written several books, the last of which is **Beyond Telepathy**.)

According to Alan Vaughn's article, Uri came out of the UFO bringing a Parker pen cartridge which belonged to Dr. Puharich (and was marked), but which Uri had de-materialized two days before. The peculiar part of the article was that it then gave no report of what Uri had seen inside the saucer. That made the whole story hard to believe, and made it difficult for me to recommend the article.

So I telephoned Dr. Puharich and asked him for details. He started by insisting that there was no objective proof of the incident, so he does not regard it as scientific. He then answered my question by explaining that Uri was in a kind of a trance at the time, and at first, remembered nothing of the experience. Later, however, he was able to recall entering a room with lots of technical equipment, lit by a yellow light with no point source. There he was handed the cartridge by someone he could not see. Again, Dr. Puharich stressed that the evidence was poor, and that he was well aware of delusions and hallucinationa, and that his only evidence for the reality of the experience was the return of the cartridge.

Many other phenomena reported in the article were much more objectively observed and photographed, and the article is very much worth reading.

In the interview in Psychic, Uri told of the possibility of much greater contact with extraterrestrials that may come about in the next three years, if we are ready for it.

This struck me as interesting because when I had attended the convention of the Universal Party in Arizona, about a year ago, members of the group who claimed to be in telepathic contact with the "space brothers" (or people who supposedly operate the UFO's), told me that they could not hope to win the presidency in 1972, but that by 1976 there would be such an intervention by the space brothers that their candidate would very likely win.

At the time, I did not give it much credence, and I still wonder, but it does seem interesting that Uri should say something similar—if we are ready.

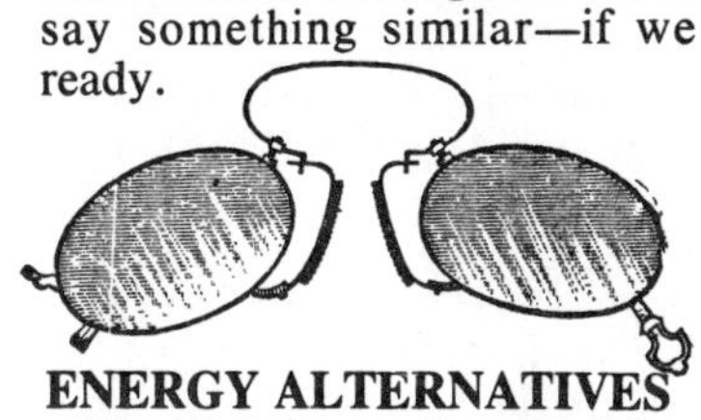

ENERGY ALTERNATIVES

The Wall Street Journal (March 6, 1974) published an article by James P. Gannon in which he found that the cost of doing an energy crash program by 1980 or even 1986 could go far too high in its effect on the economy and its disastrous effect on the environment via radiation and pollution.

One could, of course, wonder at the motives of the Wall Street Journal in taking this stand, and what high finance may be behind it, but the most striking point in our minds is the fact that in all that the article only considers oil, coal, and atomic power as its energy sources and completely omits any mention of the clean energy possibilities which could easily be developed by 1986 if a crash program in that direction would be instituted. Since those energies are in themselves totally clean (wind, sunlight, solar energy indirectly from Gulf Stream heat or special roofing that converts heat to freon pressure for Rankin type engines, etc.), there could at least be no pollution risk attached to their development.

We would like to suggest that a member of the American Association for the Advancement of Science or other groups call for a national conference to openly consider all energy possibilities rather than just the dirty ones already considered.

At the same time, it might be well to reconsider the whole of the industrial system which has given rise to this tremendous need for energy, and consider if there might not be superior alternatives which would be psychologically more suited to the true nature of man.

STANFORD RESEARCH ON GELLER: Uri Geller was tested at the Stanford Research Institute (Menlo Park, California) in November and December of 1972. A number of experiments were carried out under laboratory conditions, including tests of his clairvoyant abilities, ESP, and his ability to affect metal objects. Fantastic rumors emerged, but no scientific report was ever released. This November, however, I was able to see the statement to the press which was released, and also a 33-minute color film of the actual laboratory experiments as they were carried out. The film was shown with only a few days' notice at the Annenberg School of Communications by Jack Schwartz of the Delaware Valley Society of Parapsychology. (Jack Schwartz, 1812 Delancey Street, Philadelphia, Pennsylvania 19103 Tel. 215-732-7168). About 200 people saw it.

In both the film and the press release, it was clearly indicated that Geller was able to succeed at both telepathic and clairvoyant experiments without a single failure, and the research scientists calculated the odds against this happening by chance alone as better than 1 million to one in one experiment and better than a trillion to one for another series of tests. The film has not been released to the public and does not seem easy to come by, but I suppose the press release can be obtained by writing or phoning the names given on the release: Ronald I. Deutsch (415) 326-6200, ext. 3754 (home 326-8773) or Elizabeth de Atley at ext. 3450 at SRI.

DAWNING OF NEW CONSCIOUSNESS: "SUNSEED" is a very moving film that intends to array, in natural setting, many of the spiritual teachers and their followers who are having an impact on America. The movie may be more than just a look. A friend with me who, presumably knew nothing of the idea of photographs able to capture an image of the subject's aura, reported seeing glowing light around the heads of some of the gurus in the film. If there is a factual reality to this, the viewing of "Sunseed" may be a potent experience on more than the conscious level.

To arrange for showing of "Sunseed" in your area contact: Sunseed, c/o East-West Journal, 29 Farnsworth Street, Boston, Massachusetts 02210 (617) 482-9274/75/78.

Communities

The Magic of Findhorn Gardens

By Paul Hawken

There have been stories in the press and other media about a small community of people in the north of Scotland who talk to plants with amazing results; stories of vegetable and flower gardens animated by angelic forms where Pan's pipes are heard in the wind through the leaves; stories of plants performing incredible feats of growth and endurance: 40 lb. heads of cabbage, eight-foot delphiniums, and roses blooming in the snow, all a short distance from the Arctic circle; Don Juan and Tolkien combined where the elemental world of plants and animals cooperate with fairies, elves and gnomes in creating a land where nothing is impossible and legends are reborn; people heard talking to plants and angels in a casual and informal way creating a Garden of Eden where only gorse bushes and spiky grass grew before; a cold, wildblown peninsula jutting into the North Sea with soil as sandy and worthless as your local beach. With those sketchy and incredible tales I left for Scotland to search out Findhorn Gardens. What I found instead was something infinitely larger than 40 lb. cabbages and equally more incredible. Fairies and elves seem tame compared to what is happening here. Findhorn may very well be a manifestation of a light and a power which is so strong and divine that it could very well transform the planet we inhabit—within our lifetime. Born a skeptic, I can well appreciate that some of what I will write here will seem incredible. Neither the people of Findhorn or myself ask that you believe this account. As Peter Caddy, the founder and custodian of Findhorn, has said: "You cannot really describe Findhorn; people must experience it themselves in order to understand it."

From the window of the train I could see dark green patches of cabbages and turnips amidst larger fields of oats and barley. Soon the train would be arriving in Inverness, 20 miles from Findhorn. The heather along the tracks had turned a reddish brown. Off in the disance a large field of stubble had been set ablaze. In the mist and greyness against the orange fire stood a lone dark figure. Silhouetted with a fork in his hand, he stared motionless at the fire. Another autumn had come and once again like his father before him he had burned his fields to clear them of straw. What was the farmer thinking as he brooded and gazed into his ancestral heritage vanishing in smoke? What would he think if he knew that his fields were populated by nature spirits and angels? The farmer stood there so utterly alone, completely isolated by his act, pondering his fate as a poor crofter scratching an existence from a rocky hillside, bound by convenience to burn his fields because he was helpless to do anything else; bound by an ancient ignorance and the limitations of his consciousness.

The village of Findhorn is at the end of a narrow sandy peninsula extending into the Moray Firth. It gave its name to a weather phenomena called the Findhorn Gap. No one has satisfactorily come up with an explanation for it as yet, but basically it means that this small area of northern Scotland is unusually warm and fair considering the weather around it and its latitude. The peninsula reaches out from the Laigh of Moray, a rich alluvial plain composed of a friable reddish sandstone. There have been times when all of Scotland was hungry with famine except the folk of the Laigh of Moray. Legend has it that in the winter of

Continued →

RESEARCH QUESTIONS—ARF: John Johanning, who is presently working on **The American Book of the Dead**, has asked me to include the following note in this newsletter:

In connection with two projects I am working on at ARF, I would like to ask if any readers have any information on these subjects:

1. Has anyone tried the method of recording "spirit voices" and gotten any result? I would appreciate information to be used in **The American Book of the Dead**, which I am completing.

2. I am also interested in developing methods which people can use to help themselves develop psychic functions. I have seen the following books (recently published): **Passages**, by Mariane S. Anderson and Louis M. Savary (Harper and Row), **Mind Games** by Robert Masters and Jean Houston (Viking), **Getting there without drugs** (Viking). I have not had any experience with any groups using these books, except for a group of mind control graduates that has been using some material from **Mind Games.** Has anyone had any experience which would show how effective these books are for people who have not had preliminary training such as mind control, and are there other books or methods that anyone can recommend? —Jon.

ACADEMY OF PARAPSYCHOLOGY AND MEDICINE is the name of a scientific group which has been exploring all phases of healing in the psychic area, and they recently published a transcript of their "Interdisciplinary Symposium of October 30, 1971." It is published under the title **The Varieties of Healing Experience.** A copy of this transcript can be had for $5.00 from the APM, 314A Second Street, Los Angeles, California 94022.

The contents cover: Healing and your I.Q.—the Relationship of the elements of spiritual mind healing by Raymond K. Lilley, D.D.

Biofeedback for mind-body self-regulation healing and creativity by Elmer Green, Ph.D. (He is director of research of the Menninger Foundation in Topeka, Kansas.)

The work of the Brazilian Healer, Arigo, by Andrija Puharich, M.D.

Radionics, radiesthesia and physics by William A. Tiller, Ph.D. (chairman of the Department of Materials Science of Stanford University).

Russian parapsychology and the

the great famines in the 17th century, the people of the Laigh would put out big steaming kettles of porridge for the hungry pilgrims from the west and south.

The rich soil meant abundant harvests bringing prosperity to the region. Findhorn village, a fishing town at first, became a port of no small importance trading mainly with Holland and France. Boats built in Findhorn of native oak and larch would ply the North Sea laden with malt, salmon, and hides and returning with wines, spices, tobacco, silk, and muslin. Heavy rains sweeping down the Findhorn River into the Bay twice washed its village away and out to sea. Yet Findhorn was rebuilt and remained a port serving four-masted square-rigged schooners until the latter part of the 19th century. Now Findhorn Bay is silted in part due to that lonely crofter and his neighbors who have burned their fields so many times that the soil runs off them in the winter rains. Today, Findhorn is once again a sleepy little village and there is only one boat-builder remaining. His name is Frank and his uncle, now over 90, is the last man alive to have sailed those big schooners. Frank builds small fishing boats in his shop along the Bay of Findhorn. Short and stocky, without the typical Scottish reserve, Frank will gladly show you his shop and spend an afternoon telling tales of the Firth and the boats and men that sail it. He might even share a "wee bit o' a dramme if ye be likin'." He has a briar pipe rooted in his mouth and the twinkling eyes of a man who has seen and experienced so many grand things, that his life seems full and complete. A better weatherman you'll never find; he can predict with uncanny accuracy, and any sailor leaving the Bay for the Firth would be foolish to go against his advice.

Not all in Findhorn village are as open and friendly as Frank. There are some villagers who are full of suspicion and doubts about the "myserious" goings on of their neighbors in Findhorn Garden, a mile down the road. One lady said "some say they are a shower of weirdies. I don't know anything about that but I just don't want to go up there and get involved with whatever is going on." And another villager said that "They are well educated people. That's what maybe frightens the people. They'll know a way of influencing the kiddies." But Frank the Boatbuilder, who has spent time with the community members and who takes them sailing on his boat, has this to say about the people of Findhorn Gardens: "Greatest people you'll ever meet, Paul. They live and let live. They live in complete harmony with each other and that is obvious isn't? This is a wee village you know. You hear ol' sorts of stories; you hear a rumor here, a tale there, but it's a bit like passin' stories around ya' know. They get twisted around. I was hearing once about some bathin' in the nude and all that. Why, for goodness sakes, I do it meself off me boat sometimes. But I din't do it here off the blinkin' jetty front of the townfolk you know, and either do they now do they? At the beginnin' we heard lots of them stories but less and less now. I take 'em as I find 'em Paul. You don't go through life stickin' to what the others say now do you? There's a tremendous amount of skill up there and we get along well we do and that tis the most important thing in life! Aye, i' tis."

Frank met Peter Caddy, the founder of the Findhorn community, fifteen years ago when Peter was the manager of the Cluny Hill Hotel, the area's top resort hotel. "I knew Caddy years ago; he was managin' this place up in Cluny. Sometimes I would go there with me sailing parties for a bit of a dram or so and he would come out and greet us. He was a great guy then I thought. Always cheerful and ready with a good word you know. Very kind."

Prior to coming to Cluny Hill Hotel, Peter Caddy had been in the Royal Air Force (RAF) for fifteen years, eventually assuming the title of Squadron Leader and the responsibility of catering for the largest front of World War II in the Burma campaign. He was able to travel extensively in the Orient during those times and made many contacts and friends who would play parts in his life later.

When he got back home in the early fifties, he and his wife Eileen had an experience which changed the course of their lives. They were travelling in Glastonbury on their way to a conference when they stopped in a small sanctuary for a moment of quiet. This was a trying period of their lives when both were under unusual mental and emotional strain. In the sanctuary she heard a voice which simply said "be still and know that I am God." A still and small voice spoke to her as clearly as we speak to each other. It was the first time in her life that she had ever experienced such a phenomena. Raised in the rather orthodox Church of England, she was not prepared for this but accepted it, though not without some initial doubt and questioning. From that moment on, she was never without that "still and small voice" and she has received messages "guidance" ever since. From that point forward, everything Peter and Eileen did was with this guidance. Nothing was done without confirmation and they gradually learned to place their lives in the hands of God, following that guidance no matter how absurd or difficult it might have sounded at the moment of reception.

Under God's guidance, Peter accepted the appointment to Cluny as manager and during his stay there it

Continued →

Cayce concepts of healing by William A. McGarey, M.D.

All of the above articles are surely well worth reading.

ELIZABETH'S VISIT TO FINDHORN Shortly after Elizabeth met Peter Caddy, she visited the Findhorn Community for three weeks. When she returned, she reported on it to us.

She observed the community in Scotland as an extremely positive assemblage of people. There was an immense amount of cooperation, including the visitors.

She found the flower gardens imaginatively beauteous. The entire community is extremely clean. There are approximatley 150 members and frequently 20 to 30 guests. She added that of the members, about 50 are young Americans. Elizabeth mentioned that a substantial amount of money is earned for the community by selling plants from the garden, weaving, and making pottery.

She also remarked that the "law of manifestation" holds a high position in determining how the community's needs are met. Everything from greenhouses to tape recorders to special people are provided miraculously.

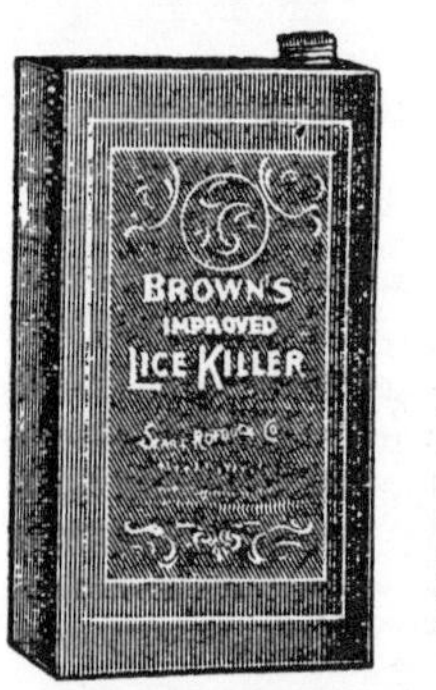

ARTERIOSCLEROSIS & EDTA: I also reported of the address of Dr. Rudolph Alsleben, who has been using a chemical known as EDTA to cure hardening of the arteries in hundreds of patients. I visited Dr. Alsleben in California in June and reported on his work. As I recall, the usual charge for the Chelation treatment (10 two-hour sessions of intravenous administration on 10 consecutive days) was about $500. Recently, I talked to a scientist and found that he had studied under the man who developed EDTA. I then learned that it is used to cure lead poisoning in children (removes the lead), and that it is not at all expensive. It can be administered orally, intramuscularly, or intravenously. Enough to treat a child costs less than $15.00.

COMIC CHRISTMAS

Has ARF gone down the drain, or made a new breakthrough in the history of science education? Our Christmas present to the world this December 1973, aside from a report on a remarkable breakthrough in controlling cancer, is a comic book! We have just completed the preliminary edition (about 50 copies, only) of the first of a series of New Age comics. This one is on plant telepathy, and touches everything from the work of Cleve Backster to the life of nature spirits in Findhorn. The picture part is followed by factual notes, which can help the reader whose imagination has been aroused to look into the factual scientific research behind the story. It owes a great debt to **The Secret Life of Plants** by Thompkins and Bird, though we had decided to do this particular work before we saw that book.

The visit of Chris Bird to us here was a great encouragement, however, and we do hope he will be pleased with the comic.

Copies of this preliminary edition are being sent to people in the publishing field, and others who would be interested and able to return it with suggested improvements, or who could help with getting it published or distributed to the public.

A CHRISTMAS FOR CANCER VICTIMS: Since the beginning of 1972, I have taken time for the study of

Soon after, communication was received from another being called the "Landscape Angel". This angel seems to oversee one specific area and reflects the thoughts of the individual "Devas" as a group. One of the important messages received from the Landscape Angel was that one must learn to think in terms of radiations when dealing with the plant world, instead of separate aspects of it such as chemicals and elements. They were told that the most important thing they could do was to radiate from within a sense of love and appreciation to the plants and that in this way every single person at Findhorn could participate in the creation of the garden.

Peter Caddy, a man who had never sown a seed before, immersed himself in the garden and amazing results began to manifest. There was the now famous cabbage weighing in at 42 lbs. with its neighbor coming in at 38 lbs. There was a broccoli plant that was almost too heavy to carry when it was finally removed from the garden after feeding the three families for several weeks. There was a continuous and amazing display of fecundity and growth throughout this early garden, all on sandy ground with barely an inch or two of soil and grass.

During this time a county agricultural adviser came and offered to do a soil analysis. Even before leaving he tried to point out that they had deficiencies, since the local sandy soils always required sulphate of potash. When he returned, it was an astonished agent holding an analysis which not only showed no deficiencies but a complete listing of all trace elements available to the plants. Because of the results that were being acheived, Peter and Eileen were offered a house with a walled garden containing rich soil. But Eileen received guidance that they were to remain on their sandy little plot and continue to demonstrate to others the power and wholeness of God.

But things did not always go right. At one time in 1967, after the garden had expanded into flowers, shrubs and trees, Peter asked a new member of the community to cut back some gorse bushes which were crowding the new orchard. Against his better judgment, this new member did so and immediately there was an uproar. Dorothy MacLean became incensed at the "butchery", particularly since the gorse were in full bloom. R. Ogilvie Crombie (See box, following page) came up immediately from Edinburgh and asked what they ahd done to upset the nature spirits so very much.

During that day, R. Ogilvie Crombie (ROC) walked towards the beach and passed through a heavy growth of gorse. As he looked behind him he noticed that he was being followed by elves looking quite distraught. They explained to ROC that their homes were in the blossoms of the gorse and scotch broom and that they were wondering why their home had been so thoughtlessly ruined. ROC stopped and spoke with them in a gentle manner and told them that man (Peter) had acted in complete ignorance. He apologized to them and they in turn said that they would return to the garden and continue to work with Peter.

Continued →

the disorder that the orthodox medical profession of this nation refers to as cancer. I have not been involved with complicated studies on a highly technical level, but have instead tried to get a wide ranging view of all the different ideas and suggestions that are being proposed from other sources, and put them together. New information from the cytology section of the National Cancer Institute is so striking that I felt that I could not wait any longer! That new information has to do with the use of hydrazine sulfate, which has proved dramatically effective, even in far advanced cases of terminal cancer.

That information was sent to me by Dean Burk of the National Cancer Institute in Bethesda, Maryland. He has been the only well-known person in that organization who has spoken out openly in favor of Laetrile, and who has publically condemned the officials who have prevented the use of that non-toxic cancer controlling agent.

When I heard on the radio that the FDA was trying to stop the sale of a form of apricot kernels that provide a natural source of amygdalin (Laetrile), I telephoned Dr. Burk. Dr. Burk then sent me a large envelope of material containing several reports on the experimental use of Hydrazine Sulfate (N2H4:H2So4), which seems to be the most promising agent for the control of cancer that I have yet heard of. He included a paper by Dr. Joseph Gold of Syracuse, who developed this treatment, and I will summarize the whole of that paper later in this newsletter.

CANCER NOT A DISEASE: Before that, I want to report some general conclusions on the nature of cancer, and offer a suggestion as to why it has been such a difficult medical problem for so long.

The system of medicine which is accepted by the American establishment is known as allopathy, and it is just one of many different medical systems. It is based on the administration of agents which produce effects opposite to the effects produced by the disease. Other systems of healing not recognized by the AMA, etc. include such systems as homeopathy, chiropractic, naturopathy, and many others.

Allopathy, in seeing cancer as a specific disease, has attempted to counteract its symptoms rather than find out what is causing a particular person to be in poor health and improving the person's health. If a way could be found to dramatically improve the health of the sufferer, then in most cases, the person's own natural resistance would get rid of the cancer. That is apparently what most of the successful, but less orthodox treatments for cancer accomplish.

In the two years of searching, I have run across a great number of these "cures," and there are so many that (according to the allopathic M.D.'s) seem to have no relevance to the disease, that all who espouse such methods are referred to as quacks.

My own conclusion is that the main cause of cancer is the combination of all the elements in our society which work together to bring about the poor health of the people. The best way to overcome the disorder is to radically improve every area of the life of patients, such as their nutrition, mental and emotional states, and often a lack of meaning in life.

But that is, of course, a big order, and hardly possible for most people who are dying of cancer. Therefore, it is important to find a way to at least control the cancer while these other changes are being worked on.

LAETRILE OR B-17: One of the oldest and best known of the controls for cancer is a substance known as amygdalin, which is sold under the name of Laetrile, or vitamin B-17. It is a natural food substance, most easily obtained from bitter almonds, but the FDA has prohibited the importing of those almonds, which are easily available in Europe. Dr. Burke provided me with an address from which German amygdalin can be obtained. I checked on that and found that the price is $2.00 per gram, or $50.00 for 100 tablets of 250 milligrams each. One to four tablets before each meal would be the usual dose. More might cause stomach disorder. For more information on that, contact Wynn Davis, P.O. Box 1206, Chula Vista, California 92012; (714) 427-3060. An alternative source of that information (which I have not verified) may be Frank Spolnick in Hammond, Indiana. Tel. (219) 933-9618.

Amygdalin can be given in far larger doses intravenously than by mouth, and is not poisonous to healthy cells, but it does seem to poison cancer cells. However, it is not a "cure" for cancer because it does not stop the body from producing more cancer cells. It is effective in many types of the disorder, but least effective in bone cancer.

Bone cancer is not readily controlled by amygdalin because the material

was raised from a three-star to four-star hotel, the highest rating. Guests would return year after year and it was always booked to capacity during the holiday season. During those years, Peter would often drive out to Findhorn village for sailing outings. A mile from the village, on what was sand dunes, marran grass, and rabbit holes, was a small caravan (trailer) park. Driving by the caravan park to the village, he would often gesture expansively to a guest or his wife and say "imagine living on a dump like that, cheek by jowl next to one another." Imagine indeed.

After six years, a neighboring hotel, called the "graveyard of managers," offered him the post of sorting it out and reinstilling life into it. But taking a hotel that had been run down and restoring it to prominence was an expensive proposition, and after the first year the owners felt that too much money was being expended, so Peter was asked to leave. So there they were, without a job and without any money, so they sought "guidance". The guidance they received was to go out to that "rubbish dump" of a caravan park and plant a garden. "Of all the foolish things" thought Peter, but by then Peter and Eileen had grown to have absolute faith in that guidance and so they moved out there. They moved into a tiny, cramped, plywood caravan which was literally next to a rubbish dump, along with their longtime friends Dorothy MacLean and Lena Lamont as well as their three children.

This was November 1962 and by the following spring they felt that they should begin to improve their surroundings. A wooden snow fence was erected and within the square that it enclosed they planned a small garden of a few salad vegetables to enhance their meals. They had no money to spend on the garden so that they had to make do with what could be scrounged from the immediate area. Soon a heap of rotting grass was found, horse manure was offered from a nearby stable, and seaweed was gathered from the beaches. When manure ran out at the stable, they went into the fields and scooped it right off the ground. A shop in town which sold produce gave them its spoiled vegetables and another neighbor donated some grass cuttings. A bale of straw found on the road was donated by a man who also contributed the salvaged wood he had bought from a dismantled building. So they were able to build and make fences and frames for the garden.

As the garden began to take shape, Dorothy MacLean discovered that she could receive messages or thoughts from certain divine aspects of the plant world. These they called the "Devas" and from them they began to receive considerable advice which was truly important, since none of them had ever planted a garden before and they were beginning on the most difficult of all possible places. Eileen received guidance that they were not to grow any flowers that first year but to concentrate on every vegetable, fruit, or herb that would grown there. During the first year, they grew 65 different vegetables, 21 different fruits and 42 different herbs. Dorothy would often contact the individual "Deva" for each type of plant and welcome it to the garden and ask for guidance.

Continued →

is not carried to the bone that well. For information on control of bone cancer, the best contact is the doctor who developed a low cost therapeutic agent known asCabasil, and he is Dr. Stuart Kabnick. He can be reached at 1405 Locust Street, Philadelphia, Pennsylvania 19102, or by telephone at (215) PE5-8330 for advice and information. Cabasil is not very expensive, is nonpoisonous, and will sometimes be supplied free of charge to people in financial difficulties.

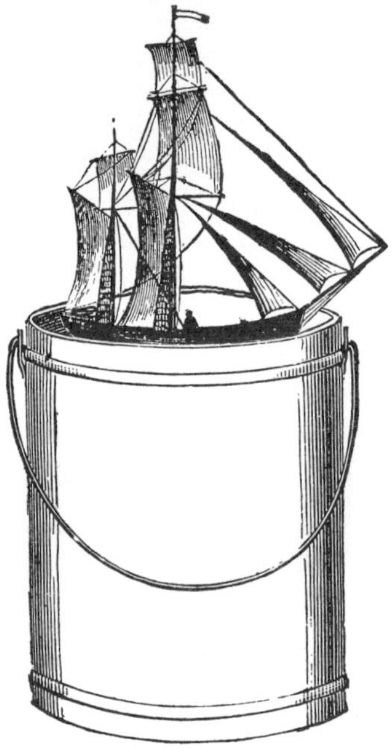

HYDRAZINE SULFATE: This is the latest, and possibly the best control for cancer which, though still in an experimental stage, has been shown to be of benefit to 60 to 75% of cases, frequently leading to a recovery of "terminal case." The information is based on a paper authored for the most part by Dr. Joseph Gold, MD., of the Syracuse Cancer Research Institute, Inc., 600 E. Genesee Street, Syracuse, New York 13202. (Tel. No.: 315 472-6616.) Though the paper was read at meetings of the German Society for Medical Tumor Therapy, Baden Baden, Germany, Nov. 3 & 4, 1973, Dr. Gold told me that it was not read by himself. Dr. Gold also told me that the use of Hydrazine Sulfate was still in experimental stages, and he did not wish to have the whole paper itself published at this time. Nevertheless, it is also clear that Hydrazine Sulfate is not a poisonous substance and we can see no reason why people who wish to use it in cases of cancer where no other remedy is available shoud not be free to do so in order to save the life of themselves or a loved one. We do not feel that this information should be withheld from people who desperately need it just because it may take the medical establishment some years to decide that it should be made freely available.

Furthermore, Hydrazine Sulfate is a relatively low-cost industrial chemical which cannot be patented, so that the pharmaceutical industry cannot make a great deal of money from it. Therefore, they may not have sufficient financial incentive to go through all the elaborate red tape procedures which the FDA requires to approve the general use of the chemical for cases of cancer. Hydrazine was the rocket fuel which the Germans used to propel the V-2 rockets in the Second World War. It is also used today for soldering aluminum with ordinary solder. The compound Hydrazine Sulfate is therefore very cheap. We have obtained some and have made it available without charge to veterinarians so that they could save the lives of pet dogs who would otherwise have died of cancer, making their owners very unhappy. Of course, no prescription is needed when the material is used for that purpose. Here is how it was found to be good for cancer:

A few years ago, it was discovered that cancerous tumors take away a lot of energy from a person because the lactic acid which the tumor produces enters the blood and is taken up by the liver, where it is recycled into a type of glucose at the expenditure of relatively large amounts of energy by the sick person's body; that it was through this process that the tumor obtained the kind of sugar it needed to live on.

The researchers then realized that if a way were found to prevent the liver from manufacturing this kind of sugar by interfering with the enzymes that are used in the process, then it might be possible to starve out the tumor without starving out the rest of the body. Various substances that could accomplish this in a live body were tested, and these were substances which were not poisonous to the body or even to the cancer cells themselves. One of the best was Hydrazine Sulfate.

In April, 1973 a 33 year-old woman who appeared to be dying of Hodgekin's disease (a form of cancer) was given Hydrazine Sulfate after she could not be helped by all the other methods of therapy that had been tried. She had already been treated by a specialist in Hodgekin's disease, and it seemed that she had only a few days

From that point Peter learned that he must be sensitive to the needs of the plant and that nothing was to be done without considering them first. Today at Findhorn you will see the gardeners talking to certain plants or weeds and explaining to them why it is that they are to be cut back or in some cases removed. This is done in advance so as to provide adequate preparation for the plant to surrender itself.

By this time ROC had encountered the god Pan on several occasions (see Box). He met Pan again during a subsequent visit to Findhorn in a very wild section of the Garden. Pan told ROC that in all gardens there must be a wild place which would serve as a sanctuary and refuge for the "nature spirits" and that it was vitally important not to disturb these places. He asked that people try to stay out of this part, and today if you happen to walk upon the right path you can find this "wild section" at Findhorn and Pan's wishes are respected to this day.

R. Ogilvie Crombie is one of the important members of Findhorn although he does not actually live there. He is in his seventies and makes his home in Edinburgh in a small flat which is jammed with seven thousand books. When he was young, he studied physics, chemistry, and mathematics at Edinburgh University. All during this time and later in life his inclination was towards science and he cherished his schooling, as it "developed the analytical side of my mind as well as the ability for objective observation and logical reasoning which makes me question everything that happens and try to find an explanation of the why and the how." ROC is a very soft-spoken and gentle man who speaks very clearly and precisely, knowing full well that much of what he has to say strains the credulity of most listeners.

Dorothy MacLean who had initially established contact with plant spirits began to be confused by the proliferation of these spirits who were influencing the life and growth around them. Up to then they had been in contact with the Devas, Nature Spirits, the Landscape Angel, Pan, and some elves and fauns. To assist her, the Devas explained themselves in this manner: "Everything manifest is in our care and we know its state, for under God we make plain his creation . . . How can we tell you how we dance with joy, that the sun dances and the moon dances and we dance around them? The earth itself dances at heart and is beginning to dance at more than heart as man rises up out of limitation and turns his eyes back towards reality . . . As the rain falls on the just and the unjust, we help to produce food for the good and the evil; the moral side of things is not our concern. We merely follow our destinies. We would say to you with one voice that our focused attention is on mankind and is clearer than in the recent past, and more precise in definition, because you are beginning to respond and be aware of our activities. At this we rejoice and are glad, for the connections between us are greater than you realize and it seems strange to us that you should be so unconscious of so much of your own being. We see you rather like an iceberg with seven-eighths of your consciousness submerged and only an eighth alive . . . We are in line of consciousness from the one great whole to the smallest unit of life and therefore we perform the miracles which you see in the growth of the seed . . ."

Continued →

to live because her lungs were severely damaged by the disease. She had lost a lot of weight and had had no appetite for several weeks. In less than two hours after she was given Hydrazine Sulfate the woman sat up in bed, said she was hungry and began to eat. Her mood and condition quickly improved, and within a few days, she had gained seven pounds. The patient continued for six months in a much improved condition but then died of an infection.

A second patient whose Hodgekin's disease was not yet considered terminal had also become resistant to all the known therapeutic treatments. He was given Hydrazine Sulfate, and also gained about six pounds within a few days. He was then able to go back to work almost symptom-free, and has remained in this condition ever since.

Those two cases led researchers to try the chemical on other terminal cancer patients with different types of cancer, and it was found that many of the patients showed improvement which could be seen objectively as well as by the fact that the patient said they felt better. Typical of some of the improvements was that the patient stopped losing weight and started to gain, appetite returned, and the patient felt stronger. Also the tumors frequently began to diminish in size, and sometimes disappear altogether.

In all probability, since the object was to test the chemical, rather than to cure the patient, I would suspect that other things such as improving the patient's nutritional or emotional situation, together with the administration of the Hydrazine Sulfate, were not attempted. If it were, more of the patients would probably have recovered more readily, but it would not have been possible to prove that the cure was brought about entirely by the use of the drug, and the value of the drug itself would not thereby be well established. What the researchers don't seem to realize is that the patient did not have cancer because of a lack of Hydrazine Sulfate, but had the cancer because he was in generally poor health for other reasons. To only starve out the cancer, but not deal with the general health of the patient would, I suppose, only result in the patient getting sick and probably dying from something else, or in the cancer returning and no longer wanting to respond to the treatment.

The fact that many of the patients did recover completely, however, may have been due also to emotional changes which the doctors had not observed. For instance, it may sometimes occur that a person who knows he has a terminal illness changes his mental outlook to the point of giving up his ideas of wealth, or his power to control the lives of his children or sometimes just willing to forgive people against whom he has born a grudge for a long time. Just that change in emotional outlook could be the thing which allows the body to function in a healthier way, raising the body's natural resistance to cancer and other disease and allowing a cure to take place which was previously not possible.

By November of 1973, it was estimated that more than two thousand cancer cases in the United States alone were being treated with Hydrazine Sulfate. Since the compound is not one that kills cancer cells as such, it does not have the poisonous side effects that most therapeutic chemicals have. The toxicity of this chemical is, in fact, quite low. Stomach trouble was found to develop in less than 2% of the patients treated (the drug is given by mouth as a small tablet). When stomach trouble did develop, it could be relieved by stopping the medication for a few days or lowering the dose. One case was reported in which the patient who had been treated with long-term steriods developed an adrenal insufficiency, but it was not known if that was due to the Hydrazine Sulfate. In several very advanced cases there have been reports of sleepiness or torpor, which was relieved by stopping the medication.

The drug is usually given in the form of small tablets, each of which contains 60 milligrams of Hydrazine Sulfate. At first, the patient is given one tablet a day for about three days. Then two tablets a day are given for three days, and then three tablets (or possibly four, if the doctor wishes) are given each day. The tablets are given between meals rather than with meals. Common scheduling is: one tablet daily at 3:00 P.M., two tablets at 10:00 A.M. and 3:00 P.M., 3 tablets would be given (one at each time, of course) at 7 A.M., 3 P.M., and 10 P.M. The types of cancer most frequently helped by this chemical are cancer of the rectum and colon, ovaries, lungs, (or bronchial tubes), prostate, Hodgekin's disease and other lymphomas, cancer of the thyroid gland, melanoma, breast cancer, and other less common types.

Obviously, those who are helped the most are those whose cancers are in the earliest stage. In early cancer,

The Devas would correspond to our idea of an architect. They seem to be the guiding force which gives form and structure and energy to the plant world. They are considered to be angelic beings who carry within them the patterns of life. This is a little different than the Nature Spirits. These we can more readily identify with, for they have apparently chosen to make themselves more manifest to man's consciousness throughout the ages for we have many stories, tales, and legends about the nature Spirits. They consist of elementals, nymphs, dryads, fauns, elves, gnomes, fairies and other creatures. They vary in size from a fraction of an inch to elves three and four feet tall. They express great joy and delight in their work, for their's is the work of actually building the plant world. They are the channel for the etheric forces which they mold and guide into the plant according to the patterns established by the Devas. To ROC they appear "formless and tenuous, like colored luminous clouds of many different and beautiful colors which keep constantly changing. They often have myriads of fine flowing lines, frequently golden which shape themselves into changing patterns. They are unbelievably beautiful and almost impossible to depict, our material colors not being pure enough."

All the Nature Spirits are under the god Pan as he is their lord. Pan described himself to ROC thus: "I am a servant of Almighty God and I and my subjects are willing to come to the aid of mankind in spite of the way he has treated us and abused nature, if he affirms belief in us and asks for our help."

Peter Caddy no longer tends to the Gardens of Findhorn. He had become so busy with the overall activities of coordination in the community that he turned over the task to a group two years ago. This was a group from Blackpool known affectionately here as the Gnomes. They were all apprentice gardeners to begin with and have, over the two years, absorbed from Peter much of what he learned during the previous nine years. They have continued to expand the Garden, as the community has expanded. One day, Gnome Leonard took me about showing me how the gardens were and are created. There is nothing used on the soil except compost which they make themselves from whatever is available. A new part of the garden is usually strrted with annuals, biennials, and herbaceous plants. This is in order to provide cover on the ground so that later on shrubs and eventually trees can take their place, the herbaceous and shrub plants providing protection for the trees in their early stages.

It was felt here that the vegetables had become a myth so to speak and that many people were simply coming to Findhorn to see large cabbages. There was a certain tendency to get hung up with the phenomena of it and to see size as an end in itself. The vegetable garden today is strong and healthy to be sure but certainly not extraordinary in the proportions of the vegetables. Matthew, the Head Gnome, a quick and prankish self-described ex-motorcycle-thug from Blackpool, prances about the Garden and community as the jester and clown, making sure that no one takes himself too seriously. He is not the sort of person you first think

Continued →

where other treatments have not been given, researchers have found almost immediate weight gain, and reversal of the symptoms of the disease, and sometimes a total disappearance of the tumor. Hydrazine Sulfate may be used together with other anti-cancer agents, and this is recommended if the other agents have been shown to be of value to the patient. Otherwise, the other agents should be discontinued and the Hydrazine Sulfate used by itself.

The report then goes on to tell of a number of cases in which this treatment has helped dramatically, but I had the impression from Dr. Gold that it would not be good to go on making great claims for the method as that could create false hopes, or prevent someone who could be treated, possibly more successfully, by another method from using those other methods. It should be sufficient here to tell you how this chemical can be obtained so that those people who want to experiment (only on their pets, of course, if they are not medical doctors) can obtain it. The chemical is available on the industrial chemical market. One hundred grams of it can be bought for $5.00 in the A.C.S. reagent form (chemically pure) from Fisher Scientific Co., 191 S. Gulph Road, King of Prussia, Pennsylvania 19406. If you buy it in a five hundred gram package (about one pound), the cost is $15.19. At that price enough of the chemical to make the equivalent of 100 tablets of 60 milligram each (normal dose) would cost 18 cents. That would last about a month if a person was taking it each day. Of course to do that, one would have to know just how to dilute the chemical with some inert substance so that the correct dosage could be given, and the little pet would not get a stomach ache. In the event that properly made up tablets each containing 60 milligrams of the substance were desired, they could be obtained for around fifty times as much as the actual cost of the chemical, and still it is very cheap (you have to remember the cost of printing labels, making tablets with marks on them, mailing in packages, etc.). To find out how to get those tablets you can call Wynn Davis in Chula Vista, California (714) 427-3060.

For physicians who wish to use Hydrazine Sulfate on patients who have cancer, they can probably get an information sheet from Dr. Gold's group in Syracuse. Of course, they will have to fill out the proper FDA forms and become part of a research study, which probably means that they can only give the chemical to some of their cancer patients, and not to others in order to see which ones recover so as to determine the efficacy of the drug. They must also comply with all pertinent federal regulations and the form to fill out is FD Form 1572. One other important point should be mentioned. Hydrazine Sulfate works by preventing the liver from producing just that kind of sugar which cancers require, so it may not be suitable for patients whose livers are damaged or who have what they call hepatic functional embarassment (liver problems).

For a published medical report on the use of Hydrazine the reference is: Abstract 332, page 83, Vol. 15, Proceedings of the American Association for Cancer Research, March 30, 1974. That article has been read to me on the phone and seems to contain less information that is given above, but may be valuable to the physician who wants to check up on the factuality of the above report. The chemical formula for Hydrazine Sulfate is N2H4:H2SO4.

HYDRAZINE CAN'T CURE: As any careful reading of all the above will show, Dr. Gold seems to have developed a very effective control for cancer of many types. As he himself says, it is not a cure. Even those whose symptoms have entirely disappeared will most likely develop new symptoms, or new diseases in the near future unless they make very significant changes in their way of living very soon. We are not proposing any particular regimen for people with cancer, but it is well to consider the difference between a "normal" American diet and that of places where cancer is extremely rare, and people are more healthy. There are a large number of organizations and books on diet and health, and **Prevention** magazine is a good source of information. One could also mention the Kelly Research Foundation, P.O. Box 89, Grapevine, Texas 76051, and their book, **One Answer to Cancer**, as a source of information which is quite specific. Another excellent book for a low cost, healthy diet is **Love Your Body** by Viktoras Kulvinskas, M.S. It is published by the Hippocrates Health Institute, 25 Exeter Street, Boston, Massachusetts 02116.

DETECTION AND CURE: The best real cure for cancer seems to be to detect its presence in the body long before any physical signs or symptoms

of when you think "gardener". At first he seems to lack the sensitivity and subtlety needed to deal with the plant world, but when you talk to him you can feel the depth to which he has become one with the garden. His animation and prankishness seems more elfin than human and when he speaks about that which he loves, he drops his voice and becomes serious and quieter: "I've seen a lot of gardens with a tremendous amount of size and uniformity, with an overwhelmin' sort of quantity of color, flash and bedazzle. But this garden seems to have a radiance. I'm very aware of a glow, a vibration comin' from the flower. Our flowers seem to say 'hiya'. It is a feeling of great joy, of love and this tremendous response back from the vegetable kingdom."

We were sitting in a rockery, built in front of the college around a lily pond. It was a sunny day with a strong wind coming out of the southwest. The Head Gnome leaned down from time to time and put a bit of sweet thyme in his mouth between pauses in his thoughts. It was just the tiniest little plant growing between the cracks in the footpath. But it was not an accident; Matthew himself had planted it in the spring and watched it thrive there. I asked Matthew why the plants were once so big: "The reason they were so outlandish in size originally was purely to demonstrate to Peter Caddy, to demonstrate that it was possible. Now we know it's possible and we can produce a plant where it won't grow normally—but that's not a new age teachin'. Just because it is out there growin' in the middle of the cold or in a dry place doesn't mean to say it is happy doing so—oh no! But even this year we had some tremendous and amazin' big things and granted it's getting less and less. We show people 40 lb. cabbages and they get hung up in the 40 lb. cabbage trip and forget. All they see is size, not the beauty."

Yet the garden is still a remarkable display of horticulture. The outsider walking through will be astounded whether he be an expert or a laymen. Paths are lined with dwarf marigolds, allysum, lobelia, pansies, snap dragons, nemecia, wooly apply mint, and columbine. They surround a profusion of petunias, livingston daisies, silver dust, asters, michaelmas daisies, poppies, companula, fuschia, delphinium, clarkia, helichrysum, monkshood, cosmos, ajuga, echium, and many others. There was hardly a single flower in the last week of September that wasn't in full bloom with more on their way; and this at a latitude further north than Moscow and almost as far north as Juneau, Alaska. The colors were as Matthew said: "radiant and sizzlin'." You can spend hours in the garden slowly moving around the caravans and bungalows amidst the beds of flowers. Behind the annuals are shrubs including hypericum, honeysuckle, flowering currants, witch hazel, wigelia, buddlea, scotch broom, lilacs, spirea, climbing roses holly, viburnum, snowberry tamarix, heathers, and some dwarf conifers.

Trees are starting to really take root now and shoot up despite the 60 and 70 mph winds that often occur here. There are mountain pine, balsam poplar, scot's and lodgepole pine, Norway spruce, cypress, douglas fir, larches, sitka spruces, horse chestnuts, taburnum, dogwood, golden

Continued →

begin to develop, and this is possible. The Kelly Foundation has urine tests for cancer and they can send you information about them.

In New York City, at the Strang Clinic for Preventive Medicine, also called the Preventive Medicine Institute (55 E. 34th St., New York), many people (mostly of the "executive" class) are examined each year for the early detection of cancer and other diseases.

Along with many other tests, each patient is required to give a sample of his handwriting, always written with an ordinary ink type fountain pen. This sample is then taken to the basement library of the clinic, where it is examined by an old Austrian by the name of Alfred Kanfer, who has worked for Strang about ten years.

He looks at the handwriting with a 50 power projection microscope, and he examines the variations in the thickness of the strokes. With an accuracy of between 80 percent to 90 percent, he is able to detect people with current malignancies, as well as those who are going to develop malignant disease as much as five years into the future.

I visited Mr. Kanfer early in 1972 after reading a long article about his work by himself and a medical doctor, which appeared in the "Bulletin of the Hospital for Joint Diseases" (New York City) in April 1958 ($3.00).

At that time, I had the idea of testing large numbers of people in this way by mail, and it seemed as if Mr. Kanfer was open to cooperate. Then it was discovered that in the contract which he has made with the clinic, he is expressly prohibited from giving the test to anyone except clinic patients. However, the methodology of the test is no secret, and no one can be prevented from using it in a non-medical manner.

The above photo is that of the EMA-4 Electromagnetic Association Motor. According to the claims of its inventors, this motor will be the power source of the future because it is not only completely free of pollution, but also is free of any need for fuel of any kind other than air. That is correct; I said **air**!

Up to now, all internal and external combustion engines or motors use some kind of liquid or solid fuel, which is burned in some manner to produce power. Most of the energy from the fuel is wasted as heat, which heats the air, and most of the energy which is actually used for mechanical purposes also finally ends up as heat.

Cities, in which most of our engines operate, are therefore several degrees warmer than the surrounding countryside, even at times when buildings are not heated. Some of that heat is due to the use of fuel in engines.

The energy taken from the fuel is not lost, but becomes heat energy in the air. If a way could be found to convert that heat back into mechanical energy in an efficient manner, we might finally have a non-polluting engine that runs without any fuel at all except for the heat in the air. Warm air would go in and cooler air would come out, while a great deal of power would be produced.

If my information is correct, such an engine has now been developed in California. I have been invited to come and see it myself.

I am also aware that it is rather unusual for me to tell about a new discovery without telling all the details as to where the engine can be seen. The reason for that is that the inventors are saying that they are being overrun with "wall to wall people," and are asking that no further publicity be given to the development at this time. Also, most of the people with whom I have spoken tend to disbelieve that such a fuelless engine has now been developed, and I don't want to make too powerful a claim until I have personally seen and tested the machine.

MAN'S PRESUMPTUOUS BRAIN by A. Simeons, M.D. Dutton pb 1961 $1.95.

A brilliant study of psychosomatic disorders by an extensively traveled specialist. It is Dr. Simeon's contention that man's brain battles with and submerges his animal instincts, and that this battle grows more serious with every cultural advance. The result of these "resumptions" of the conscious mind over the involuntary body systems and their needs can easily manifest as disorders of the upper and lower digestive tract, of the heart and blood vessels, the thyroid and metabolism, and some forms of diabetes, arthiritis, and sexual deviations.

His discriptions of the mechanisms behind psychosomatic disease are the best we have ever seen, based as they

elder, mountain ash, birch, hawthorne, and four brave eucalyptus trees which have not been seen in these parts before and technically cannot grow here. Some trees have grown so tall so fast that the gardeners and community have asked them to stop so that they root deeper and protect themselves before the powerful winds.

Experts who have come to visit Findhorn are astonished by the rich and healthy variety of plants all growing to perfection in the lowest of quality of soils. In certain areas are windbreaks of soft berries which, as of the first week in October, were heavily laden with fruit after having fed a community of over a one hundred throughout the summer. There were hardly any weeds at all anywhere—they just do not seem to grow here. I saw no pests while I was there and the only insects I did see were butterflies and bees. Bumblebees would land on the cedum in the morning and by afternoon they would be scattered on the ground around the flowers, drunk and woozy with nectar. I never saw a single leaf eaten by a green bug or worms or scab of any type. Everything looks absolutely healthy and vigorous. Professor R. Lindsay Robb, a consultant to the Soil Association who has been appointed to agricultural schools and missions throughout the world as a soil expert, wrote that "the vigour, health, and bloom of the plants in this garden at mid-winter on land which is almost barren powdery sand cannot be explained by the moderate dressings of compost, nor indeed by the application of any known cultural methods of organic husbandry. There are other factors and these are vital ones!" When one first sees Findhorn Gardens, the riot of color almost overwhelms your vision and perceptions. But time allows the eye to rove in the corners of the gardens and study more closely the plants. Picking up the soil with one's hands is startling. You go down a couple of inches and you are right back to sand. Yet staring at you in the eye and waving brightly in the afternoon breeze is a compact and profuse group of the brightest and healthiest flowers you have ever encountered. Everything here defies rational analysis. You can chase your tail trying to figure out what is happening here at Findhorn in the plant kingdom. There is no explanation that scientific authorities can offer to explain the phenomena of Findhorn. Experts are baffled and so will you be unless you listen to the stories of Eileen, Dorothy, and R. Ogilvie Crombie.

The gardeners today do not directly perceive the Nature Spirits or Pan but rather act on an intuitive level, listening to that inner part of them which guides them through the day and allows them to become the instrument of the garden. However, Matthew spoke of Pan saying that "Pan is me closest and most treasured friend on Earth . . . I mean . . . there is no separaton between me and Pan. I move through this garden and . . . quite often . . . I am totally one with Pan. And so are you! Just like one can be one with the Christ, one can be one with Pan. There's no separation at all. You cannot do anything without the knowledge of Pan. You can't breathe without his consent. Pan is the body within every life force, everything that

Continued →

are, on clearly demostrable observations and scientific methodology. The writing is brilliant, intended for lay people, but many help to enlighten many doctors too.

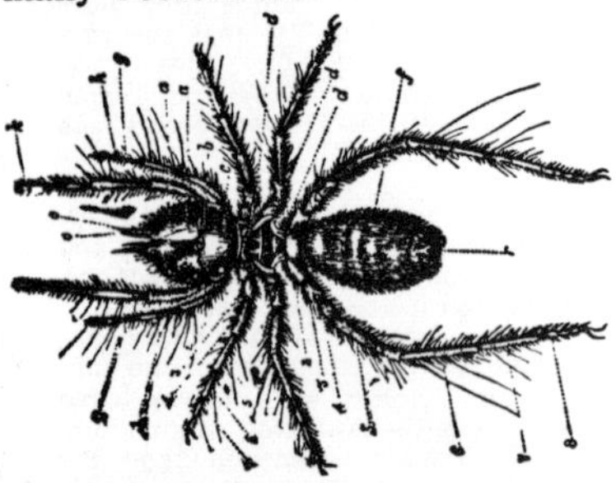

PUPPIES PUPPIES PUPPIES: April 13, Laurie (Lady's daughter) has given birth to ten very healthy puppies. They are most black German shepherds from a most exceptional mother and a good father. We have raised Laurie from a pup and she is a brilliant creature with very loving, open-minded attitudes—pups will probably be as special.

IS THERE AN ENERGY CRISIS? Is the energy crisis real, or is it just another Watergate? Is it another attempt of the President to gain power over the lives of Americans, together with the oil companies with whom he has always cooperated? The only facts that we or the government seem to have are the ones released by the oil companies—except for a very good article by Murray Bookchin in the December 20, 1973 issue of **WIN** magazine.

Evidently, there are excellent sources of power such as solar energy, wind and even the considerable temperature difference of the Gulf Stream, which are not polluting or a radiation hazard such as is the case with measures the President has proposed. This **WIN** magazine can be had by sending 25 cents to **WIN** magazine, Box 547, Rifton, New York 12471. There are also some other very worthwhile articles in the same magazine. Get it!

IS THERE AN ALTERNATIVE PRESS? Yes, there still is an alternative press in the United States, and if you would like to have a free list of the names and addresses of 200 alternative papers presently publishing, just send a stamped, self-addressed envelope to: (the list includes foreign papers): Alternative Press Syndicate, Box 26, Village Station, New York, New York 10014. (Tel: 212 242-3888)

MORE ON ELECTROMAGNETIC ASSOCIATION MOTOR: A magazine article first turned me on to the motor. The article was obviously written by someone who was not a scientist, and who did not understand how a motor could run without fuel. Nor did he seem to understand the tremendous social, economic and political implications of such an invention.

It was the kind of article that looked real, but like so many things that never get beyond the planning stage, most readers would expect never to hear of it again—just one of those fantastic inventions that are fun to talk about but which don't really pan out. And if it was real, we reasoned, why did it not appear as headlines in the **New York Times**?

But since my work often involves a willingness to check out the ridiculous, I decided to try to telephone the inventor, and soon found myself in touch with his right hand man. He seemed extremely open and friendly, welcoming scientific inquiry, yet not sounding too scientific himself. He did not, however, sound like a con man.

So I called up one of the best scientists I knew in this area; a man who has a deep spiritual understanding, concerned for peace, ecology and psychic research, and who happens to be the inventor of the Bell helicopter. (We all know the invention but few know the man who struggled for 18 years to make it a reality).

My friend was very skeptical. He'd heard of perpetual motion before, and told me it was a waste of time. But if I insisted and had to bring it out to him, he would be willing to have a look at it the next morning.

After studying the articles and specifications, he was not convinced. I let him hear recordings of phone calls

procreates, everything that dies, everything that moves. Pan as popularly understood is a distortion, just like Jesus. People put Jesus on a cross, and people put Pan to hoof and horn. I think it is a beautiful analogy of Pan, a very beautiful picture of the intellect of man on top and an animal lower. Aaaahhh, in some of me attunements with Pan I feel this tremendous procreative force, a dynamic urge, this surging forward! . . . Yet on the other hand, being with Pan is a bit like lace curtains and glistening gossamers . . . You see a world slightly out of focus . . . It sees the little grasses, the little ferns, and the mosses growin' in between the cracks. This too is Pan. This little sweet thyme down here that I'm nibblin' is every much a part of Pan as two elephants, and the little worm in the sea is as much a part of Pan as the great Blue Whale. Pan! . . . He's so vast and dynamic and yet so sincere and gentle. These painters that paint the woodlands . . . You know well that they, knowingly or unknowingly, have been gently kicked by Pan. This tremendous knowing is just like the Christ energy that causes you to help an old lady across the street . . . or build a light center . . . It is all the same thing."

In this case the light center is Findhorn and that is how it is referred to by the members. Perhaps in that expression rests more of what Findhorn is and how it manifests this incredible creative and spiritual growth than that which is implied in the word "garden". For the surprise at Findhorn is not the plants, although they do grow in amazing and wondrous ways. The absolute wonder of Findhorn is the people it grows. There is at Findhorn an aura of harmony and love amongst people, for the lessons of the garden have been carried over to the world of beings. Within Findhorn is a community of people who have tapped the secret of group consciousness, of sublimating the ego to the whole, of letting go of the desires of the self. Actually this sounds quite ordinary because isn't that what you read about everywhere and is that not what so many are now trying to do? What makes Findhorn singular in this respect is that talk is the least of it and action and dynamics are the largest. It is called a "working" community and this has two meanings. It means not only that it is a community which is functioning within a new consciousness but it means that it is succeeding as well. The people of Findhorn believe that the way to the New Age of man is through seeking and following God's guidance. And they do simply that. They seek it and follow it. There is no belief here that God's guidance is limited to a few or that they are "chosen" in that sense. Findhorn is a place where everyone is encouraged to find that voice within him or herself. What makes this successful at Findhorn is that it began with people who had developed that ability and as more and more people came, the love and light which led these people was shared with the newcomers allowing them to grow similarly and thereby allowing the community to grow.

Although this community grows, it does not go out and work for anything it requires. Everything here is produced by the "law of manifestation" which is the tenet that if you are following that voice within you which

Continued →

with the inventor. The matter began to get interesting and a few minutes later he was helping me out with some cash to fix up the '66 Dodge and pay for the phone calls to California. By then I was mentally prepared to go to California and see the real thing if it indeed existed.

I had already written to my brother, who is a California physicist, and had asked the inventor of the motor to send him some information. I sent him a tape of the phone call. I then called my brother on the phone and asked him to have a look at the machine to see if it were an obvious fraud. I was assured that he could tell that it was without even reading the articles, and he wasn't willing to spend the money or time to go and see. I said I'd pay his expenses if it was a fraud; he offered to give me $10,000 if it were real. Finally, he agreed to read the articles, hear the tape, and telephone the inventor on or by May 22. Then he would call me back to tell me if he was willing to visit. I await his call but after several more talks with the inventor, I feel sure enough myself to make the trip, unless he sees it and discovers a fraud that cannot be denied.

EVGRAY ENTERPRISES AND EMA MOTOR: After I had spoken with the inventor, Edwin V. Gray, I decided that I had to leave immediately for California. He had promised me that I would be able to see the motor run under load numerous times, without the batteries needing a charge.

I had spoken many times with Richard Hackenberger, Ed's assistant, and understood that I was entirely welcome to see and test the machine. We arrived in California on May 25, and I went to see the motor the next day. Before seeing the engine, I was required to sign a paper that I would not disclose anything of the principle of operation that I might observe.

As soon as I saw the motor, it seemed apparent that it could not get much energy from the air because the air inlets and outlets were too small to move much air. The building also did not look like a research laboratory or workshop. No tools, etc.

Instead of being permitted to test it in some way, I was given demonstrations of the "principle of operation" by being shown how a truck battery could be used to cause a one pound electromagnet to jump 18" into the air with a big bang, and was then told that the battery would never run down because the "energy was being recycled". But there was no evidence given. Also, no indication that any energy was obtained from the air. The demonstration was dramatic-but insignificant.

In a second demonstration, an ordinary-looking motorcycle battery was used to light six 15 watt lamps and some other units totaling less than 200 watts. I was told that the lamps would not get hot, and the battery which could only deliver 2 ampere hours would not run down because the energy was being recycled. I tested that and found that the lamps get just about as hot as 15 watt lamps usually do, and when I checked with Harley-Davidson Motorcycles, I discovered that the battery supplied ten times the power that they were claiming.

It could have run the demonstration for well over an hour, and they only used it for about ten minutes.

To find a sure way of testing the motor, I consulted one of the top engineers of the GM Proving Grounds in Van Nuys, California, and together we came to the simple proposal of discharging the batteries, running the motor, and see if it recharges its batteries while running from them. The motor could not be endangered by such a test.

When I tried to arrange that, I spoke with EvGray's general manager (G. DeMos) for over an hour, but he would not agree to any test at all, or to my reentering their building. He knew I suspected a fraud.

Thereupon I contacted the Los Angeles County District Attorney, learned that he was already investigating. I gave him a five-page report. I also visited Dick Hennesy, editor of **Probe**. **Probe** had run two articles on the EvGray device. Since they claim to publish only true stories, they should be asked to do a thorough investigation or print a retraction.

I also visited Larry Liebman of **Southern California Industrial News**, which had run a favorable article. He seemed unwilling to publish the results of my study for fear of legal action.

OTHER NEW MOTORS AND VEHICLES: Jim Bush of the GM Proving Grounds is the man who looks over hundreds of ideas for new types of engines, and he told me of an article in **Car and Driver** magazine for June

---

is the higher consciousness common to all men, then you are "in the right place, at the right time, doing the right thing" and all of your needs are met. Your needs—not your desires. Life at Findhorn therefore calls for a surrender which is at once frightening and ecstatic. It is simply that one must give up his or her will to that greater will which we suppress and ignore in favor of our own desires and self. The people surrender themselves completely to this guidance and the will of God. The faith that they feel is like a rock—it is immovable. Such absolute faith can have its problems when it occasionally meets the dualistic consciousness of the confused beings who stalk the earth, and I guess that is just about everybody including me. People, sometimes a little self-righteous in their ambivalent state, are sometimes shocked by the absolutes of Findhorn. This is not a community where itinerants can wander in and by invoking their presence claim "rights". There is no minority view here simply because there is no majority. There are not two sides to an argument because there are no arguments. What sounds wonderful to some may sound a bit frightening to others. At Findhorn, there are no regulations, there are no orders, there are no chains of command, but there is a group constantly striving to maintain and improve their receptivity to God and each other in order to channel light and truth to Earth. So there is no rule or knowing how something will come to the community and likewise there is no planning for the "future"—there is only the simple faith that in time all needs will be met. Up to now, they have been met 100% and there does not seem to be any sign of letting up. The love and tremendous positive feeling that emanates out of Findhorn has a remarkable affect on people—even those who have not visited it, so it is not uncommon to receive anonymous donations of money or other items.

This "law of manifestation", as the keystone to the material development of Findhorn, functions like prayer, except that first the message or guidance is heard to do or build something. Nothing is just asked for. Then, when that message is received, sometimes as a feeling by Peter or sometimes in meditation by Eileen, they 'spec' it out. If it is a fairly reasonable project like outfitting an office, Peter will often go out and buy it with no money in the bank. Yet during its twelve years they have not written any overdrafts, and have paid every single bill directly when due. The money always appears from somewhere when needed. If it is a large project like the college building, they will simply wait and concentrate on it. There is no constant prayer, for they feel that it is a sign of weakness reflecting little faith. They have an absolute faith that after such guidance—i.e. "you will have the building on the hill for a college"—that it will come and soon too. This has only once failed. Guidance was received that they would soon have a greenhouse. Peter so requested and waited. Nothing happened. For over a year nothing happened and it began to puzzle them, for this was unheard of. Again they received guidance that they would receive a greenhouse. Then Peter measured the space and made a precise and specific request for an eight-by-twelve cedarwood greenhouse and it arrived within the week. Unbelievable? Maybe, but after listening to

Continued →

1973, on an electric car developed in Pennsylvania.

The problem with electric cars is principally the batteries; presently marketed batteries do not carry enough charge per pound to make them practical for a car. However, Jim thought that batteries have already been developed which can carry ten times the charge of the ordinary battery. These are suitable for electric cars.

Batteries would probably be provided on lease basis, so that the car would not have to wait for a recharge, but could pick up charged batteries in a minute.

One correspondent pointed out that an electric car would pollute as much as an internal combustion (gasoline) car because of the terrible pollution involved in electric power plants. That is true at present, but I believe that it is far easier to build non-polluting stationary plants than vehicles; we could have had them long ago if people had cared enough.

THE MINTO ENGINE: Back in 1971, I heard about Wallace Minto's freon vapor engine, which was successfully demonstrated in Florida. The patent sold to Datsun of Japan, who promised to build and test 100 of them the following year. They ran on kerosene without polluting.

Recently, I checked with the Datsun people and they replied: . . . "We think it is only fair to say that the Minto engine is still basically in the research stage. Major problems such as size, weight, cost, fuel consumption, lubrication, and durability still lack acceptable solutions. These problems are more complex due to Datsun being a smaller car than the large American models.

"For these reasons, and with a relatively short time period to solve the automotive emission problem in accordance with the Clean Air Act, we do not foresee vapor engines becoming important automotive production engines, or as being able to displace current internal combustion engines and their associated emission control packages within the next ten years."

When I compare that letter to the detailed discussion between Mr. Minto and Senator Thomas Eagleton (March 14, 1972), I get the impression that the reason Datsun is not producing a non-polluting car is very different from that which is given in their letter.

The full report can be had from: Kinetics, 1121 Lewis Avenue, Sarasota, Florida 33577.

Why doesn't General Motors do something to produce non-polluting cars? I raised that question with Jim, and the answer I got was that GM was still paying off the cost of going into production on the V-8 engine, and that in spite of the fact that it only costs $35 to produce a new V-8 engine. The reasons then appear more financial than scientific.

DO-IT-YOURSELF Pollution reduction: For those who don't want to wait, and who have $100 or $200 to lay out, there is a solution that is worth considering. You convert your car to run on propane. To get all the dope on that, write to: Jerry Friedberg, P.O. Box 531, Point Arean, California 95468.

He can supply all parts and information at very low cost, and propane has 70% less pollution, no lead, and makes the car last longer; run for less. I'm thinking of it too.

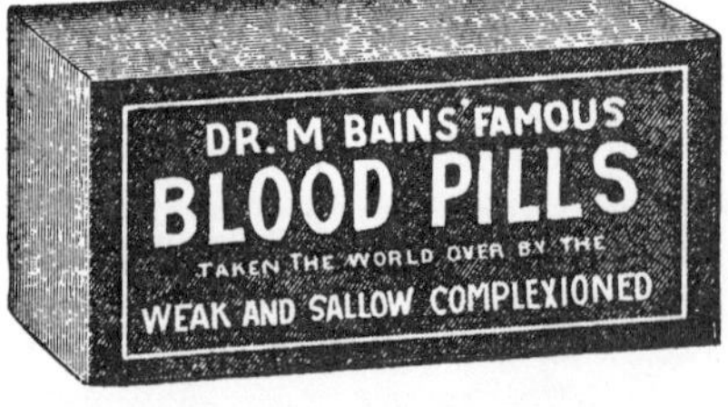

UNDERGROUND AND PREVENTIVE MEDICINE

While in California, I had the opportunity to spend some time with Dr. H. Rudolph Alsleben. The appointment he had made with me by phone weeks before was the first one after his return from an out-of-town trip; perhaps that was a reason why he did not have as much time for me as I would have liked to spend with him. We can get to the point at once, and I explained that I would like to learn about his work by experiencing it, but did not have much money.

His immediate impression was that I was undernourished, and he sent me to his nutritionist, Ethyl Davis, for an interview. She agreed that I needed better nourishment, and gave me some suggestions after learning about the

Peter Caddy recounting how everything here "arrived", "showed up", "manifested", was "donated" or "given", you can only accept and marvel at his faith and your doubt. From a plywood caravan set astride a trash dump 11 years ago to the abundant and prosperous community of 180 today lay a thousand incidences of the "law of manifestation" which Peter will gladly and joyfully recount to you by the hour with great humor and gusto. All this when people give themselves to God. "Give and thou shalt receive" is an active principle here, operating on both the material and spiritual levels. It will make anyone stop and think a bit as he wanders about the community looking at the new cedarwood bungalows, the double caravans, printing presses, darkrooms, a grocery store, a craft shop, a "Park" and college, a supermodern kitchen and dining room, greenhouses, a theater, workshops, a Sanctuary, and the exquisite Gardens of Findhorn.

And after touring the community with Peter and listening to him recount the saga of Findhorn, one gets to be terribly curious about this man who seems to be the seed of all of this development. He appears perfectly ordinary in most ways. When I first arrived in the commnity, he came bustling out to greet me. A ruddy, powerfully built man in his mid-fifties, he has those sparkling eyes that grandfathers get when they feel the wholeness of their life in their latter years. Peter, however, is vigourously young in feeling and energy. He is lightning quick in thought, speech, and action. There is never a moment of hesitation or doubt in his words. He is like a force unto itself, whirling about the caravans and offices. And he is the most unlikely person to be the "custodian" and founder of a "New Age" community. There is nothing guru-like or "holy" in Peter at first sight. He is not the soft-spoken master gathering the children to his feet. He rather makes everyone stand on their toes, catalyzing the energy daily in a dynamic way. He seems very straight, middle class in fact. But then the art of perception is to learn that appearances and reality are not the same but quite different and you will never encounter another person where this is so remarkably true. Outwardly straight, English, and very much the Squadron Leader barking out commands, inwardly Peter is a man for which we have no words yet in English. The Sufis always caution the seeker to look not for the obvious signs when seeking a teacher and there are countless stories of how ragged, haggle-toothed beggars wandering out of deserts were nothing but saints and sages. This is the twentieth century. How about a clean cut, fiftyish, ex-RAF officer with three children and a lovely silver-permed wife living in a trailer park with a garden of delphiniums and marigolds?

Impeccable in speech and manner, Peter speaks to you without noticeable accent in the clipped style of the military. But then this allows him to make himself perfectly clear. He has a large store of aphorisms and cliches which everyone here knows by heart. He constantly re-emphasizes these to all around him in a clear and concise way. In this way the working principles of Findhorn are constantly brought out and the lessons sink in. To the many

Continued →

diet I had at present. I will try to reproduce those suggestions at a later time.

Dr. Alsleben then took a thermograph of my face. That is a photograph of the face in which the slightest variations in skin temperature can be clearly seen (as little as 1/2° is obvious). From that, he explained, he is able to tell if arteries that serve the head have any blockages, even if the blockage is only partial. The test indicated some blockage, which is common.

At the time, it did occur to me that the mind also controls skin temperature, and I wondered about that later. It has been shown at the Menninger clinic that some people can vary skin temperature as much as 10° by mental control alone.

The greatest surprise came when a blood test was made. The doctor took a drop of blood from my finger and looked at it under his dark field microscope.

DARK FIELD MICROSCOPY:

What we saw was amazing to both of us. Most surprising was that in spite of my apparent "undernourished" condition, my blood was completely free of any type of bacterial infection, such as most people examined seem to have. We saw a few white corpuscles, which are the bodies that fight infection, but they were wandering around with nothing to do.

At the same time, we also noticed a peculiar condition of the red corpuscles. Although they appeared to be round and healthy, all of them were "sludged" together to such an extent that there was not a single free red cell to be seen. The doctor made a photograph of this condition.

I would have liked to spend some time talking with Dr. Alsleben about these things, but it was apparent that he was extremely busy with other patients whose need was much greater than mine. Also, he told me that his book, which had just been published, would be available in a few days, and that he would send me a copy. I therefore felt that I should read that before taking any more of his time. I did, however, spend the next hour or two looking at blood samples through both the dark field and ordinary microscope, and I was amazed at how much more could be seen through the dark field method than the conventional bright field system.

A DARK FIELD microscope is actually only a conventional microscope with a different type of condenser. Whereas the regular instrument projects the light through the specimen to be seen only by the way in which it refracts or impedes the passage of light, the dark field condenser projects the light from all sides and allows you to see the specimen with reflected light, much as you see ordinary objects in the room with you. The difference is so striking that I wonder why all medical microscopes are not of that type.

Ordinary microscopes can also be used in a dark field mode (though less perfectly), simply by beaming light at the specimen from an angle above the slide instead of from below.

CHELATION THERAPY is a method of removing excess calcium (such as arteriosclerosis), lead, and other poison from the body. It is done by administering a chemical called E.D.T.A., which captures the calcium, lead and other ions. Because the E.D.T.A. is not absorbed by the body, it is excreted entirely, and the calcium, lead, mercury, etc. go out with it. The treatment is generally considered safe and harmless for most people, and is given once a day for ten to twenty days, depending on tests.

By this method, Dr. Alsleben believes he can overcome hardening (calcification) of the arteries, as well as the effects of lead in the atmosphere (result of leaded gasoline). He believes lead to be one of the major causes of many organic disorders, such as multiple sclerosis, etc. (That disease was unknown before the automobile, and is rarely found where cars are scarce—or so I'm told).

About 30 patients are chelated at a time, and three groups are done each day. Dr. Alsleben felt that I should not have the therapy at that time because I seemed to be somewhat under-

food-zombies wandering from diet to diet, the reply here might be you are what you think as well as what you eat. The stress here is on creative and positive thinking. Peter is the embodiment of this. He emanates an incredible amount of confidence and assurance, yet missing is the arrogance and pride that usually accompanies such feelings. One never feels in Peter's presence the slightest sense of selfishness or self-worth. He feels that this lifetime and past lifetimes were all training for what is happening now and what will happen at Findhorn in the future. He feels Findhorn has been successful because he has given it a very strong grounding in discipline and structure. He constantly emphasizes to all the importance of order and discipline; that freedom is only realized through discipline, not around it. So when Sanctuary starts at 8:30 in the morning it means precisely that and, after 8:30, on goes the redlight at the entrance and no one else enters until Sanctuary is over. But then people are hardly ever late here. This sense of discipline is almost totally self-imposed and a high degree of self-discipline is required in order to be a member of Findhorn community.

There are no signs here saying "don't" or "no". In fact there are hardly any signs saying anything at all. Communications here are almost totally oral and are done at group meetings. Within Findhorn there are no titled presidents, managers or officers other than the title of "focalizer". It is a recently coined term and it means that such a person focuses the energies of a group or department. The hierarchic structure is blessedly absent here. Craig Gibson focalizes the focalizers as a group. A two-year resident of Findhorn, he is Australian by birth and has spent the last twelve years wandering the globe doing everything from being an Argentine Ant Inspector, to roles of sales clerk, weighmaster, farmhand, stockman, longshoreman, lifeguard, male model, painter, sailor, skin diving instructor, carpenter, mime actor, sculptor, and finally before coming to Findhorn did a two year stint at a nearby Tibetan meditation center. During his stay at the retreat he studied Mahayana Buddhism, Hatha Yoga, and Macrobiotics He then left for Findhorn which he had heard about and upon arrival immediately fell in with the maintenance and construction crew. He then helped construct the Community Center and focalized the making of the furniture for the dining room. These are superb tables and benches made of native ash split vertically, sanded smooth, and finished off with a deep, glossy, resonant glow which reflects the warm amber lights around them. The dining room can double as a theater in the round or a dance floor and makes a warm and inviting place for the community to meet on the windy and stormy nights.

As focalizer-at-large, Craig is constantly channeling the energies of the various groups and giving of himself, his love and energy. He answers questions directly and clearly and keeps everything going at a high level. There are several meetings a day and these meetings animate the flow of community events and action. The types of meetings are numerous and ever-changing since the community is divided up into 17 departments. That

Continued →

nourished (and, I supposed, because my need for it is not especially urgent).

Dr. Wilfid Shute, who is known for his work in vitamin E as a treatment for heart and circulatory conditions, was treated by Dr. Alsleben with Chelation therapy at a time when two of his heart valves were so badly calcified that surgeons were advising an open heart operation for the replacement of the valves—an operation for which survival rates were not too good. Dr. Shute had been having blackout spells, and it was said that he could drop dead at any moment, when the treatment was undertaken. Today, Dr. Shute is well, and is living in Anaheim and working with Dr. Alsleben as part of his team. So also are nutritionists Dale Alexander, Ethyl Davis, and a number of others.

I was not there long enough to make a thorough study of the Alsleben Foundation and its work, but my immediate impression was that it was on the level, and one of the most scientifically advanced preventive medicine research projects I know of.

RUSSIA'S KULAGINA: A person with startling abilities is Madame Kulagina in Russia. Many writers have written of her abilities, but now we have a report in a reputable paper by a physicist who went to Russia to test her, and then returned months later with all the sophisticated equipment required to do scientific experiments in a place where no fraud would be possible.

This extremely interesting article by Dr. Benson Herbert, director of the Paraphysical Laboratory, Downton, Wiltshire, England (also editor of **Journal of Paraphysics**) is found in Volume 4, Number 4 of **Parapsychology Review**. Copies of that issue can be obtained for 85 cents. Address is 29 W. 57th Street, New York, New York 10019. Subscription is $4 per year.

HEALTH AND HEALING: New Light on Therapeutic Energies is the title of a book by Mark Gallert, M.D., M.Sc. In this edition (1966) Mr. Gallert discusses a very large number of healing and therapeutic methods which have been left unpublicized by most of the healing professions.

The book goes into detail on the work of Wilhelm Reich (Orgone Energy), chiropractic, homeopathy (old and new), chromotherapy (use of colored lights), and many others. All of these are brought together in a book which would be of interest to both layman and doctor. The book is published in England by James Clarke, Co., 31 Queen Anne's Gate, London S.W. 1. $5 should cover cost and postage. This book is also available from YES Books, 1039 31st Street N.W., Washington, D.C. (Tel: 202 338-7874).

There are a large number of experiments shown in the book which could be duplicated with a minimum of scientific equipment in order to verify whether or not the information is accurate. None appear to be dangerous.

The theory that different colors of light can help to cure various ailments has been heard from time to time. Many large books have been written on the subject, but most of us have assumed it to be a kind of quackery, and have not bothered to investigate. **New Light on Therapeutic Energies** devotes about 30 pages to a summary of the best (and rare) volumes on this matter.

Personally, I would not have taken the matter very seriously until recently, one correspondent directed me to EHLRI (the Environmental Health and Light Research Institute —associated with New College, 1873 Hillview Street, Sarasota, Florida 33579).

I wrote to them and was sent much printed material (free). Here was evidence based on a large amount of good scientific study, that the color of light can and does affect human health; that there is a part of the eye that has a non-visual function of influencing the endocrine system.

"This was a wild idea—really far out," recalled John Ott, director of the Institute, "that light entering the eyes influences the endocrine system, which controls basic body chemistry through production and release of hormones."

". . . The whole wall of resistance is crumbling now," said Ott, whose work through EHRLI has now received funds from several foundations, as detailed in his book **Health and Light** (published by Devin Adair; price $7.95).

Evidently, then, there may be something to chromotherapy after all.

seventeen, plus variations, makes for a rather complex yet obviously viable system by which the community constantly interweaves and embraces the energies, ideas and consciousness of some 140 permanent residents and 35-40 guests. It molds them into a group consciousness which transcends the individual and keeps the community squarely in the hands of those higher forces, those forces which cause plants to shoot forth from a barren earth with startling vigor.

At a recent "internal conference" the community was divided into six groups consisting of about 20 people to a group. Our group was focalized by Craig, and discussion proceeded easily from most in a wandering manner touching on diverse subjects but with everyone returning to the experience of loss of self, sublimation to the whole and a sense of attunement to all. Although it would appear that within this context, talk would be cheap, there existed within our group a greater measure of oneness than lip service. The Findhorn magic is one of the actual, palpable experience of one's own consciousness merging with a universal consciousness. Discussions are not philosophical debates or expositions. They are more likely to be one individual after another describing the experiencing of oneness, the seemingly infinite joy that this unfolding of higher consciousness bestows, and the lightness of mental gait when the load of being "someone" is cast away and the dance of simply being "one" begins. In the room we centered in, the feeling was so strong, that words beggared it so we merely sat in silence at times. Someone here likened the process to "witnessing," where people recount the experience of "Christ" consciousness and of giving oneself over to Christ. Well, that is what is happening alright. People here however have taken Jesus off the cross and into their hearts as a friend and brother who simply stood up and said I am the son of God and so art thou . . .

Enfolding at Findhorn is an environment highly conducive to the transformation of consciousness. *I never heard anyone at Findhorn criticize anyone else while I was there. I repeat, during my two week stay, I never heard a single negative word about another person. There is no set of dogmas diets, meditative techniques, or physical exercises to aid or bring about such consciousness.* A friend from Boston relates the story of how he met some ex-Findhornians in California recently and how he felt this great joy and happiness from them. He asked them what they were "in to" and they all stopped and looked at each other for a moment and then laughingly replied: "I don't think we're into anything."

There is no benign and powerful influence of a "holy" man shedding his light on the souls of others. People at Findhorn have told me that they have experienced the greatest feeling of freedom here at Findhorn. The lack of a strong "spiritual" leader does not seem to hinder the spiritual growth of the community at all; in fact the absence of such a person seems to positively enhance the ability of others to grow within themselves. All spiritual teachers here are invisible—the angels, the Devas, Saint Michael

Continued →

HOW TO SURVIVE THE NEW HEALTH CATASTROPHES is the title of a book by Doctors H. Rudolph Alsleben and Wilfrid Shute (of vitamin E-fame). The book of 446 pages is published by Dr. Alsleben and Shute privately, apparently for their patients, and the price is $6.95.

This volume deals with the effects of pollution as a cause of many of our modern medical problems, and with the treatment of heart disease and hardening of the arteries by vitamin E and by chelation therapy. It shows the most effective methods of actually curing those diseases, but that the medical profession is not generally open to them.

In addition, the book discusses the effect of pollution poisoning and the symptoms which are actually very common. I find that there is evidence here that my own slight problems might be due to mercury poisoning (which is indicated to be common among lithographers and photographers, and I have been into both of these for a good number of years). To determine this with certainly, the book advises laboratory tests, and I will have that done soon.

How to Survive the New Health Catastrophes can be obtained from: The Cheer Medical Center, 710 N. Euclid, Anaheim, California 92801. I may write more about the book when I have managed to read it all.

SILVA MIND CONTROL IN SCHOOLS: Recently, the Silva Mind Control course, which has been taught to the public for some years, was taught to the whole student body and many teachers of the Hallihan High School here in Philadelphia with such positive results that many other school administrators have been asking to have the course taught in their schools. In response to this, the Silva people have inaugurated a new approach in which they have agreed to have the school systems let some of its own teachers become qualified Silva instructors, capable of teaching the mind control program to school pupils.

Before publishing this, I called the Silva Institute in Laredo and spoke to Harry McKnight, who said that school superintendents, with their board of education agreement, can recommend qualified personnel to the Silva Institute for instructor training. They should do this by writing to Dr. George DeSau, director of educational research, Silva Mind Control International, P.O. Box 1149, Laredo, Texas 78040.

In the case of private schools who are not under the city board of education, the principal of the school may also write to Dr. DeSau.

PSYCHIC AND HEALING: The August 1973 of **Psychic** magazine is entirely devoted to the subject of healing, with articles on various healers and methods by capable authors. It is, in general, a scientific account. To get the August issue, send $1 to **Psychic**, P.O. Box 26289, San Francisco, California 92126.

URI GELLER: A STEP FORWARD IN UNDERSTANDING
by Jerry Shifman

If one is performing a scientific experiment and all of the "data points" except one fall into a neat, predictable pattern, there is a strong inclination to throw out the odd reading on the grounds that is must be an error. And indeed, in the majority of cases this is true. But once in a great while that bit on nonconforming data are valid, and indicates an occurence which was completely unexpected.

On these occasions the "odd reading" turns out to be the most significant information of all, leading to a new insight, a step forward in understanding. For the experimenter these rare moments can be very exciting and gratifying.

Similarly, in daily life we all tend to disregard information which is clearly absurd. When little Johnny comes in from the yard and says that a giant polkadotted elephant is flying overhead his mother is not likely to go rushing outside to check it out for herself. We all feel that we have a pretty firm grip on what is possible and what is not possible, and we simply refuse to believe things which we know to be impossible.

But how is one to respond when one is confronted with solid evidence that events which are clearly impossible

and Saint Germaine, Jesus, Pan and a host of others. All seem to give guidance here.

Every morning before anything starts, there is Sanctuary for everyone. It is in a square building with chairs arranged circularly on a carpeted floor and room for people to sit or kneel on the floor if they so choose. There is no denominational sign of any sort and the only adornment is a weaving that depicts "light, which was donated to the community. Morning Sanctuary is usually punctuated by Peter reading off the list of the day's activities and meetings. Occasionally he will read letters from friends who have left the community. Eileen will also share with everyone the "guidance" that she may have received, reading from a small red book. It is usually direct, simple, and sometimes quite eloquent. An example of a message is as follows: "Be at peace. It is the opening of the heart center which makes you feel everything so deeply. You find that tears flow easily? These are the tears of joy; let them flow. They are the tears of recognition of Truth and Love Divine. This brings a feeling of elation, of thanksgiving, of a new and deep understanding. It is as if you had slept and have now awakened to find a new world. Everything looks different and beautiful, as if you went to sleep in winter when everything lay dormant and have awakened to spring in its full glory. The same trees, the same garden are there, but vibrant with life, growth, color, and beauty, for all is made new and you feel new in the process . . . Rejoice, my child, rejoice. Tread delicately, alertly, so that you miss nothing. See even the tiniest insect in a new light. Because you have chosen to open your eyes all is indeed made new. You have divided up life, but now it is one glorious Whole. The wholeness fills your heart, expands it, raises your consciousness. This newness has to start within and then expand outwardly and up and up. There is no limit to this expansion; it can go on forever and ever."

There is another Sanctuary at night which is sometimes accompanied by a tape of the New Troubadors, a musical group formed at Findhorn, singing songs written and composed there. The group was founded by David Spangler, a 28-year-old California native. The songs have a timeless quality about them with pure and lilting melodies which echo in your mind long after you've heard them. It is a remarkably professional sound too, despite the fact that the group had little or no musical experience before coming together at Findhorn. David Spangler was born to an Engineer and his wife prematurely, weighing four pounds at birth though losing his twin brother in the process. His early life was plagued by a weak constitution and asthma but not his mind. He showed exceptional brilliance at an early age and when he was fifteen he began to give lectures on "New Age" themes. In the latter part of the sixties he tried to gather the energies of some friends towards a "New Age" community in northern California but it did not seem to get off the ground. He then read of Findhorn and the experiences of Eileen Dorothy, and ROC. He immediately sensed that this was a place and an event of enormous and timely importance: "In June 1970, I arrived at Findhorn on what was to have been a brief five-day visit. Liter-

Continued →

are actually happening? One must make a choice: either valid reasons must be found for disregarding the evidence, or one must reevaluate his understanding of how the world works.

For one who has been trained in science this can be a most uncomfortable mental bind. For this writer the experience is both exciting and very unsettling. Here is how it came about.

A SCIENCE REPORT: On March 9, Russell Targ and Harold E. Puthoff, both scientists working for Stanford Research Institute, took part in a physics colloquium at Columbia University. For the first time in public they revealed details of experiments they had performed during November and December of 1972.

The object of the experiments was "to determine whether so-called psychic or psychoenergetic functioning can be observed under rigidly controlled laboratory conditions." One of the subjects they tested was Uri Geller, an Israeli with alleged psychic powers.

In reporting the results of the experiments, SRI carefully avoided using words such as psychic, clairvoyant and telepathic. SRI stressed that the experiments had been arranged in advance by the staff scientists, and were specially designed to be "cheat-proof."

But they did not hedge at all regarding what the subject actually had been able to do. In a dry, unadorned style it was stated flatly that Geller had repeatedly performed tasks which were beyond the explanations of conventional science. For example:

A single die was closed in a solid metal box and shaken vigorously. In ten trials Geller made eight calls and was right every time. The other two times he said he couldn't tell and declined to guess. The odds against this happening by chance are better than a million to one.

A small object was placed in one of twelve identical film canisters. Without touching the cans, Geller determined which one the object was in. He did this twelve times without any errors. The odds are incredible.

There was much more, but these two examples will give you some idea of what the SRI report stated.

I GO TO SEE GELLER

My first knowledge of this came from the Berkeley **Daily Gazette**, which printed a summary of the official SRI news release. The Gazette also mentioned that Geller was going to have a public demonstration in Berkeley in April.

The next day I contacted SRI and spoke with Ron Deutsch, their public information man. Yes, it was true. Yes, Geller actually did the things printed in the Gazette. No, they had no explanation at all for how these phenomena "worked." Very curious! Needless to say, I was in the audience when Geller appeared for his public demonstration at Zellerbach Auditorium on April 14.

Geller was introduced by Andrija Puharich. Dr. Puharich is a neurologist and a researcher in paranormal phenomena. He has spent more than twenty years in this field of study, and has worked with many well known psychics, including Eileen Garrett and Peter Hurkos.

In **Beyond Telepathy**, published in 1962, he presented his highly technical speculations regarding the sort of mechanism that might be involved in telepathy. In his brief introductory talk he said that he had been studying Geller for the past eighteen months and described him, quite cordially, as a unique freak of nature.

Geller himself is a good looking fellow about twenty-five years old. He is poised and unassuming on stage, somewhat boyish and ingenuous. Part of the impact of his demonstration was due to the totally convincing air that surrounds Geller and Puharich. Nowhere in either of them is there any trace of deviousness.

30 PERCENT CORRECT, BUT NO GUESSING:

When Geller got down to actually doing psychic feats, he made it clear that he could not always make it work, and that he could only promise to do his best and see what would happen. As it turned out, he was successful in about 30% of the things he attempted to do.

An interesting point regarding this is that he is able to tell whether or not he is getting a telepathic message clearly. When it is clear, he says so and gives the correct answer with considerable confidence. When it is not clear he gives no answer at all. At no point in the evening did he "guess wrong." So, in that sense, his accuracy was 100%.

Incidentally, he used many volunteers from the audience, and he always called for women. Whenever a volunteer came forward he asked people in the audience who knew the person and could vouch for them to raise their hands. Repeatedly he asked his volunteers, "Do you go to the university? Are there people here who know you?"

ally within hours I knew that I had arrived in a place manifesting the most powerful vibrations of new direction and creativity that I had ever experienced, vibrations that were anchored and balanced in a most practical fashion. Here was living proof of communication with other realms of life. Here, too, was proof of the presence of God as a living, guiding reality; the power and vibration, the beauty and growing balance of the community were too tremendous to be denied or explained away as simply some kind of psychic manifestation. Here was a demonstration of a new way; man worshiping God, the Beloved, by joining in creative partnership and oneness with Him and working to build a bit of Heaven on Earch for His glory and the upliftment of all".

The five days stretched on to three years. David became a dynamic part of the community and began the education aspect of it which is located today in the "college". As the focalizer of the college, he wrote prodigiously, a series of books and pamphlets, gave a series of some 30 lectures which are taped and distributed throughout the community and world, and generally introduced to others the concept of Findhorn being a major transitional link to the "New Age".

Today, one has very much the same impression as David did in 1970. Today, the manifestations of a higher will are even more evident and clear than then. When I left the States for Scotland, several people asked me to report back and tell them if "anything was going on up there". Something is most definitely going on up there. That something cannot be easily named or given a handle. If you have ever visited a room where a "saint" or "holy man" has lived and meditated, you know that feeling of experiencing a presence which defies analysis though it is so easily perceived. At Findhorn, that "presence" or sense of vibration is so strong and powerful, that anyone can notice it. Simply speaking, Findhorn seems to be one of those places where an enormously positive and divine energy is anchored and transfixed. This energy does not flow from Peter Caddy but through him. And one can feel it flow through everything here, from the flowers to the children to the elderly people in their eighties. It is a transforming power that one feels and it changes all who stay here. Those who cannot accept this energy leave quickly, so there is no problem of who stays and who leaves. The vibrational energy in the community is a great challenge. It is not a place to fall back and be comfortable, although there is a certain degree of security indeed living among people who love and respect you and each other very much. You are required to constantly re-examine yourself and your actions within the context of the whole, and this process of examination seems to lead to a surrender of the lower energies which hold most of us back.

It is felt here that because we concentrate so much on our "image" of ourself, we must constantly hold ourselves in check and readjust either ourselves or external reality to conform to this image. Since this is essentially an energy turned in on itself, it does not renew itself so easily. This leads to mental and physical fatigue, self-consciousness, lack of self-assurance, and a hindered vision of true reality.

Continued →

A blackboard was placed where the audience could see it, but he could not. A volunteer would write something on the board and Geller would try to pick up what it was from the minds of the audience. First the names of two colors were put on the board. He couldn't get either one. Two different colors were written on the board. He got one of them, correctly stating that it was the last one written.

Then the name of a city was put up. No luck. Geller was clearly chagrined at his poor showing, and apologized on several occasions. Another city was written on the board (Dublin). He got this right, then asked who in the audience was repeatedly thinking, "Moscow, Moscow?" No response. Then he indicated a general area in the audience and insisted that a woman in a yellow dress had been persistently thinking "Moscow." Then a woman stood up. She said she was the one. She was wearing a long blouse that reached below her hips. It was bright yellow.

A drawing was put on the board. A volunteer drew a stylized swan. After concentrating for some minutes, Geller finally said that he had part of it, but that he couldn't recognize it. He went to the board (the board had been flipped over so that the original sketch was now on the back side, out of everyone's view) and drew three or four strikes that clearly matched the gross outline of the swan. Then he backed away and shook his head. He couldn't figure out what it was supposed to be.

He was blindfolded and a random woman was called up to the stage. He described how she was dressed.

He said that he could sometimes make broken watches run again. Several watches were passed to him. He and his corps of female volunteers handled them and focused psychic energy on them. No results at all.

BENDING A RING

He than began a series of attempts to distort or bend physical objects without touching them. His method was to have a volunteer hold the object between her palms while he placed his hands outside hers and concentrated. Several attempts were made with various objects. No results. He took a breather at this point and talked for a few minutes about his feelings about his abilities.

Following that he took a ring from a woman in the audience and once again tried to get something to happen to it. This time something began happening—the woman who was holding the ring first said she felt a tingling. Then she said it was getting warm. Then she said, "It's moving!" Finally, when she opened her hands, it was seen that the ring was bent out of its original shape. The ring was the sort that was adjustable; that is, the part that goes around the finger was slightly overlapped. After it had been Gellerized, one of these legs was bent and curved sharply inward.

That concluded the demonstration.

What is one to make of all this? First, it must be stated that this sort of public demonstration, taken alone, is certainly not solid evidence. I am sure that a halfway competent stage magician could produce similar effects through trickery. But that line of thought comes to a full halt when one considers the rigidly controlled experiments done at SRI.

That is another matter altogether. It is hard to believe that the SRI scientists could have been repeatedly fooled by simple trickery. And if the SRI results were real and not sleight-of-hand, then we must seriously consider that the events that took place at Zellerbach might be real.

After stewing over this a couple of days, I found that my thoughts kept going back to that bent ring. Somehow I felt that I could perhaps swallow the telepathic events, but that business with the ring was bizarre. If that really happened the implications were staggering. Here was the classic situation. I was confronted with an event which my scientific mind could not accept, and yet there seemed to be no valid justification for disregarding it. I could neither swallow it nor spit it out.

At this point I began contacting anyone I could find who knew Geller or was in any way associated with him. Through the organizers of the Zellerbach demonstration I was given the name of the woman from the audience who had held the ring.

Since then I have gotten to know her, and we have had several long conversations. Her name is Janina Hamm and she works as a physical therapist. Janina described to me in great detail what happened on the stage. There can be no doubt at all that she believes that Geller caused that ring to deform while she held it between her palms.

She felt it moving. Further, Janina herself is completely sincere and convincing. Whatever was going on with that ring, it certainly did not involve any duplicity on her part.

So where does this leave us? If Janina Hamm held the ring and she

Those who are able to release this heavy burden of "image" and personality experience a great release of energy which was formerly used to hide and conform oneself. Findhorn provides for people, young and old, a matrix within which they can rapidly undergo this process of transformation. The energy which is released is merged with the energies which come from higher levels. The merging of these two energies creates a synergistic effect where the whole exceeds the sum of the parts. The remarkable thing about Findhorn is that so many here are living embodiments of that change, yet Findhorn lacks obvious techniques, dogmas, or religious doctrines to hasten along this process or bring it about. It is done in a non-denominational, non-dogmatic, individual way, each in his own pace and way without external guidance.

There is therefore a tremendous creative energy at Findhorn, the energy which is latent in all but released here. It manifests itself daily in a variety of activities and these activities are carried on with great laughter, gaiety, lightness, and joy. Throughout my life I have seen somber-faced people churning out work for the sake of some cause. In fact that is usually what I have been doing. But here there is no cause, in that sense, except the expression of God's will in a practical and human way. Everywhere you go at Findhorn, people are having fun. They are having so much fun in fact you may well wonder how any work is done at all, yet this lightness of spirit belies the fact that so much is done.

The community is spotlessly clean. Guests are taken care of superbly and graciously. The gardens are a wonder to behold. In the kitchen, instead of one steady staff, there is a rotating staff of cooks and cook's helpers. Each meal then becomes a joyous creative act unto itself by the people involved. In most kitchens, the cooks become so tired that the food does not reflect real positive vibrations. Here, a new staff every day prevents that and yet it is not chaos. The kitchen runs very smoothly with good coordination provided by the focalizer. After dinner, many people spontaneously go into the kitchen and help clean up. It is accomplished quickly and there seems to be no hang-up over sexual roles as to who should do what. If anything, men seem to predominate in cleaning up the kitchen. The meals are vegetarian with dairy products and eggs. Lunch is a largely cold buffet while dinner is served to you at your table. There is always a good variety of foodstuffs so regardless of your food "leaning" you can eat a nourishing and well-balanced diet with no fuss at all. Flexibility, a word stressed throughout the kitchen and community, is practiced here with the result that you hear no talk of food except praise and compliments.

The theater aspect of Findhorn is another source of amazement. With perhaps fourteen people and a theater no bigger than some living rooms, converted out of an old tool shed, it produces hilariously funny, entertaining, and moving skits, mimes, dances, and songs. Some of it is great theater which could hold its own in any city of the world. Yet there is a message which comes through in an unaffected manner. Theater with a message is usually a disaster with the message coming through but not the "theater". It comes through here, yet

Continued →

believes it, why can't I accept it? At what point does healthy scientific skepticism become an irrational refusal to believe information which doesn't fit preconceived notions?

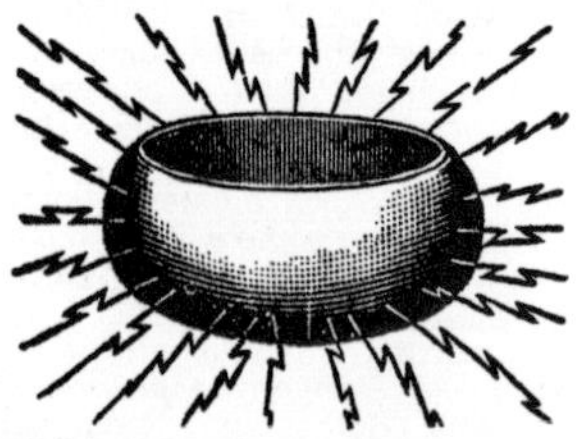

MORE ON GELLER-—PERSONAL ACCOUNT:

Last Tuesday, September 25, 1973, I was given two tickets to a public appearance in New York City of Uri Geller.

I would have liked to have a front seat, but hearing about it on short notice, I had to be content with the balcony and a powerful pair of binoculars (10X) Additionally, a closed circuit projection TV screen made it possible for everyone to see a good enlargement of what was happening.

I was most impressed right at the start by Uri's simple honesty in telling about himself and his experiences. More than any other speaker I have yet heard, I would say that if I could do the things that Uri does, I believe that I would present the matter in the same way that he has presented it. He seems simple and honest.

He started by inviting people, mostly women, to the stage, apparently without any choosing on his part. There followed a few fairly good telepathic tricks, which were not very evidential. After that, he invited each one who might have brought along a watch that didn't work to bring it to him. He got only about eight. He said that in Texas he got thirty, and was able to make eighteen go again. To do this, he had a girl hold the watch between her hands, held his hands above hers, and concentrated on it. Finally, he would clench his fist hard a few inches above her hands and then listen to the watch. None worked!

Uri apologized for that; when he first started to speak to us, he admitted that he was a bit nervous, and not because of the audience (said he liked people, that he also had no enemies —and asked that no one shoot him). He was nervous because he had no idea what might work and what might not, and he did not want the people to be disappointed.

After answering many questions in a simple and straightforward manner (except for one), he then went on to the task of psychokinesis, for which he is most famous: he would attempt to cause a piece of metal to bend, using only his mind.

He asked women to bring to the stage any strong finger rings they may want to offer for this experiment, and of the several that were brought up, he selected the strongest, made with fairly thick metal—far too thick to bend without a pliers. He asked a girl to hold it between her hands as the watches were held. He concentrated; nothing happened. He tried another girl. No luck. Than he asked if anyone could supply a still heavier ring. None came. Then he concentrated longer and harder, and the girl who held the ring said she felt something moving (it was between her hands).

The ring had now come apart and had begun to bend open. It happened slowly, and as she had now opened her hand, we could see the ring on the palm of her hand slowly bending open more and more.

I was able to feel it afterwards, and it was a strong ring; far stronger than I could bend without a pliers; and, of course, we did see it bending and saw that it was not done in any mechanical way.

During the appearance, Uri spoke freely about some of his own history, likes and dislikes, and about the negative things that **Time** and some other publications had said about him. When asked if he could outdo the professional magicians, he said he was not particularly interested in that, but if any of them could "fool the scientists" as **Time** said he had done at Stanford Research Institute, then he would be willing to take them on. The demonstration was convincing.

After it was over, many people rushed up to see the ring and get Uri's autograph. I want to greet Dr. Puharich, who looks far younger than I had expected. I found him in back, looking at a large photograph held by a man whom he seemed to know. On a previous occasion, this man had come to Uri with a camera, the lens of which was well covered with a lens cover that was taped on (so absolutely no light could enter) and Uri was asked to attempt to put a picture on the film inside the camera. When the film was developed, it showed a clear image of Uri and the people in the room, etc., all in perfect focus, except that the lens cover could be seen as a perfectly round circle with two small ears (which it had), and this blotted out most of the picture. Of course, one could suppose this to be trick photo-

does not affect the quality of performance or wit of the writing, because the people are not any different from that which they are trying to communicate. They project to the audience the attunement and harmony that went into their production so that even a cold reserved audience of upper-middle-class Scotch eventually break down in gales of laughter and murmurs of praise.

Yet these are not people who spend their days in rehearsals and elaborate preparations. They cook, make pottery, tend the store, work in the college, and many have children which they care for during the day. The creative aspect unfolds at Findhorn in a completely natural way with ease and grace. People who had done "nothing" creative before coming to Findhorn emerge as great dancers, photographers, potters, and carpenters. There are evenings in the community center given over to spontaneous entertainment. The harmony and closeness of the group creates an electric atmosphere where people feel free to express their innermost thoughts in the beauty of dance or song, individually or in a group. And always, here and with everything else that happens in the community, there is laughter and humor.

There is a studio for weaving and macrame, one for graphics, one for pottery, and one for candlemaking. These crafts are sold at Findhorn and in craftshops throughout Scotland. Again, the quality of the work is uniformly high. There are also darkrooms providing a means for several members to assemble slide shows which are accompanied synchronously by tapes of the "New Troubadors" or by talks by David Spangler or Peter Caddy. These shows are given to visitors and are also shown outside as well as being sold to other groups throughout the world.

A great deal of literature comprised of Eileen and Dorothy's "guidances" and David's writings is sent to people who request it. Along with tapes by David and ROC, there is also the Findhorn News, a quarterly, and a book called "Findhorn Garden" explaining the early history. All of it is written and printed by the members of the community on their printing presses. Festivals are celebrated four times a year at the solstices and equinoxes but one-day festivals on a more spontaneous level are not uncommon. All of these activities are an expression of the gratitude and joy that people here feel from receiving the abundant gifts of God.

At this point Findhorn might sound like certain fundamentalist communities that dot the American landscape, and exoterically there is a similarity to be sure. But esoterically, Findhorn is unfolding in a pattern which is as startling as it is vast.

England is dotted with "power points" and "etheric webs" as John Michell in his book *View Over Atlantis* brought out. These "etheric webs" are nothing but lines or fields of energy which connect individual "power points". These "power points", like Stonehenge and Glastonbury, are places where etheric energy is transformed into a more dense form, often through human consciousness. This principle is extremely important for the proper

Continued →

graphy, and it could have been, but I did not have that impression from the people there. Dr. Puharich quickly went out and called Uri in to see the picture, which he had not yet seen.

As to the one question Uri did not answer in a simple way, it was about his relationship to UFOs. He said he would like to leave that for later. Later, he came to it and told us that if he were to tell all he knew, it would be so far out that we would not believe him. That although it may sound like a sales pitch for a book, he felt that the best would be to wait until Dr. Puharich finished the book **Uri** that he is currently writing, and there we would get all the facts exactly. However, he did feel he could tell us that something quite astounding would occur (for the good) in the next few years.

Uri himself expressed a complete willingness to subject himself to any amount of scientific testing, though he admitted that at one time (in Israel), he was not interested in that, fearing that he could do nothing in a laboratory. Since then, he has been carefully tested by the Stanford Research Institute—one of the best—but the SRI scientists have not been willing to release their full report! I wonder. . .

SCIENTIFIC STUDY OF DOWSING: In their magazine, **The American Dowser**, the American Society of Dowsers, Inc. (Danville, Vermont 05828) has recently published a scientific study of field dowsing.

(Dowsing is the detection of underground water, minerals, pipes, etc. by the use of a hand-held device, such as a forked stick, etc.)

In these tests, Dr. Zaboj V. Harvalik, a physicist and scientific advisor to the U.S. Army's Advanced Material Concepts Agency, performed meticulous tests with one of the world's most sensitive dowsers, Wilheim De Boer of Bremen, West Germany. Dr. Harvalik (of Lorton, Virginia) then published five papers about these tests, and these, together with an article by Christopher Bird, are in the August 1973 issue of the magazine.

De Boer was tested for his ability to detect currents as weak as one microampere (which produced a magnetic field effect of less than one billionth of the earth's magnetic field that a compass detects).

With A.C., tests were made throughout the range from one to 75 cycles per second, and then throughout the audio frequency band, and on up to one megacycle (R.F.). Some frequencies were detected better than others, and some not detected at all.

Additional tests were then conducted using high frequency radio beams in the seven meter band and in the ultra high five centimeter band, which it was found De Boer could also detect. Then experiments were performed to determine with what part of the body these radio waves were detected. It was found that the beams were detected in the solar plexus area and in the head.

Copies of the complete reports can probably be obtained from the above address for $2. Year's subscription to the publication is $6, along with membership in the society

ASTRONAUT EDGAR MITCHELL has founded an organization called the Institute of Noetic Sciences (575 Middlefield Raod, Palo Alto, California 94301), and this group seems to be working along very similar ideas to Aquarian Research: attempting by use of scientific and psychic studies to open people's minds to the coming of the New Age and the new possibilities that this may bring.

HYPNOSIS CONFIRMS FEAR OF TWO WITH UFO STORY

Pascagoula, Miss.—(UPI)—Two men who claim they were taken aboard an unidenified flying object by three creatures with crab-like hands and pointed ears definitely underwent a "terrifying experience."

This is the conclusion of two scientists who questioned the men under hypnosis.

Dr. Allen Hynek, chairman of the astronomy department at Northwestern University, and Dr. James Harder, of the Aero-Phenomenon Research Orgnaization and the University of California, spent several hours interviewing Charles Hickson, 42, and Calvin Parker, 18.

Taken Inside Craft

The two told police they were fishing off an old pier on the Pascagoula River Thursday night when they saw a strange object approaching in the sky, emitting a bluish haze.

understanding of the possible implications of Findhorn. This etheric web is analogous to our "soul". It is in fact the soul of the Earth and without it the planet would be indeed dead and lifeless. This etheric power which may be called the world of vibration of the pre-atomic world animates all living forms. It is this energy which the Nature Spirits form into plants, and it is this energy which the Devas give structure and flow to; and it is this energy which science is still looking for.

A "power point" is where this higher energy is commingled easily with the denser forms of energy on this planet and where people who are sensitive can more easily receive "guidance" or "transmissions" from this higher level. These "power points" focus these higher energies to earth, and holy men throughout the world have located temples and churches on these very spots. If we can establish contact and gain access to this realm of vibration and energy, then a steady flow of energy can be channelled to this denser realm we occupy. This energy can be used and transmuted to influence physical and spiritual life on the planet. There are legends of using such energies to build the Pyramids and other seemingly impossible structures. There are stories in all ancient cultures of how these energies were used for the benefit of mankind.

In England these "power points" were linked on the ground by straight roads and paths so that individuals traveling throughout the Isles could stay on the track so to speak. These lines were called "ley lines" and they are still visible today in England although they are gradually being obscured. Most of the ancient "power centers" have become obscure as well, over the past two thousand years. Their source of power seems to have vanished or been diffused. The Pyramids lie cold and empty, the Mayan centers are grown over by jungle, and Stonehenge has partially collapsed. Most of them have become objects of tourism and curiosity rather than understanding and use. Taking the body again for comparison, it is as if body of the earth had become weakened and "clogged" and the strong connection between one's physical and etheric body had been partially sundered. Just as we, through our duality, have become dissociated from our higher self, so the earth has become divided and separated from its "etheric web" of divine energy which is our common birthright.

We are reaching an end of an age. Astrologically, we speak of the Piscean and Aquarian as being synonymous with the "old" and the "new". There are, as well, many books of a holy nature which point to this time of transition, not the least of which is the Book of Revelations in the Bible. And while some may agree or disagree over whether catastrophe, man-made or natural, will accompany this change, there seem to be more and more people who are tuning themselves into this vibration of change. It is felt across all lands. People who had developed ways of opening their consciousness through meditation, yoga, food, zen, christianity, chanting, and other ways seem to be awakening to the fact that the paths we took out from the world of limitations were convergent; we seem to be awakening

Continued →

Thé men, reportedly "scared to death and shaking all over," told officers they were taken inside the craft by three creatures with big eyes, crab-like hands, wrinkled skin and pointed ears. They said they were examined and released unharmed.

Harder said the men's experience as reflected during hypnosis was "traumatic," "Their emotions and very strong feelings of terror are impossible to fake under hypnosis."

Hynek, who was scientific consultant to Project Bluebook, a study concluded by the U.S. Air Force on UFO's in 1969, said, "There is no question in my mind that these men have had a very terrifying experience. Under no circumstances should they be ridiculed."

Definite Pattern

Harder said the phenomenon has happened all over the world and seems to follow a pattern.

"A UFO tale is an incredible tale told by credible persons," Harder said.

The men, both employed at the W. B. Walker Shipyards, were checked for radioactivity and the results were negative. Several persons in the Pascagoula area have said since they also saw a UFO.

Startling advances in holography were demonstrated, and theories proposed to explain both precognition and the function of the brain.

Dr. John Jungerman, professor of physics at the University of California at Davis, described an experiment in the moving of a pendulum by psychokinesis, which he is measuring in a manner similar to that of the famous Michelson-Morely experiements of 1889. He predicted a similar impact on physics today.

He also spoke of the work of Dr. Benson Herbert from England, who was able to test Russia's famous psychokinetic medium Kulagina under laboratory conditions. He also described the working and charging of Pavlita's psychotronic generators told about in **Psychic Discoveries Behind the Iron Curtain**.

The complete report of the Benson Herber testing of Kulagina is also published in Parapsychology Review, Volume 4, No. 4, which can be obtained from 85 cents from Parapsychology Foundation, 29 W. 57th Street, New York City, New York 10019. Subscription is $4 per year.

At the Prague conference there were about 3,000 scientists from 10 countries and 15 different scientific fields.

SAUCER SIGHTED BY PENNSYLVANIA POLICE On the morning of October 4, 1973, two Pennsylvania State Police wrote the following report:

"October 4, 1973: Myself, Corporal Bugjo, and Trooper Cutrufello were on routine parol. We were traveling south on route 352, approximately 2:15 (A.M.). I observed an object in the sky hovering over an area of the Penn State campus. I saw a bright light coming from the bottom of this object. The light was facing downward and I brought this to the attention of Trooper Cutrufello; we stopped the police vehicle, got out of the vehicle and watched this thing for approximately 90 seconds."

I was able to spend an hour interviewing both of these men together about six days later, and made a tape recording of all that they could tell me. They indicated that the object had red lights around it in a teardrop shape, with the front appearing to be the point of the teardrop. The light was shining down from the front quite brightly, and the white light shining down on the campus seemed strong enough to dissipate the mist that was in the area clear out of the path of the light beam. The beam spread out at an angle of about 40 degrees. After 90 seconds, it went out suddenly (no cooling effect like a filament makes), and the craft with its red lights started to move slowly to the east. After it was about a quarter of a mile away, the red lights also went out. Then they heard, for the first time, a sound, and it was like the purring of a cat coming from the vehicle (craft). That ended shortly and nothing further was observed. They then called by radio to another car which was patrolling to the east of them, and asked those men to meet them. On

to the invisible and visible signs which are accompanying the transformation to the New Age. New and strange weather patterns around the globe speak of an Earth that is coming alive despite man's continuing destructive efforts. Listen to Eileen again and what she has heard: "This is an historic and momentous time in the progress of man. At this time the veil is being rent in two and that which has been hidden through the ages is now to be revealed. The secrets of the sages will no longer be secrets, for all shall know about them. The Book of Life has now been unlocked and the truths held within it are being brought into the open for all men to see and to use. Revelation upon revelation is now being revealed. My glories are now being manifested here on earth. It is all happening now. Open your eyes and behold it. Miss nothing. All is there for all men to behold when they choose to open their eyes to do so. The time is at hand when these words will become reality: "The Son of God will appear in full glory for all to see." Be ready, for no man knows the day nor the hour. Be prepared at any moment to receive Him. I tell you to be surprised at nothing."

David Spangler writes, "The New Age is a release of new energy. It represents for the planet an unfoldment to the time in a man's life when his soul takes a tighter grip over its lower vehicles and his body, emotions and mind and all become orientated to the direction and life of the Higher Self. At such a time, the person ceases to have a personality in the sense of a veil or mask between himself and his true divine nature. The lower vehicles that make up the personality become direct channels for the manifestation of the consciousness of the soul. This, of course, releases into the body a surge of new energy and vitality, of new direction, creativity, vision, and life. This is what is happening to Earth. What could be called the 'personality level' of the planet is falling away, being transmuted; old thought forms and complexes of energy . . . corresponding to neuroses within the human psyche, are being erased, transmuted, brought to the surface and cleansed. In the place of this, new levels of energy are coming to the fore, taking charge, resulting in both a cleansing of the etheric body of the earth and the revelation of a new body . . ."

In other words, new energies are infusing the earth's soul which is beginning to regain its grip and oneness with its physical body, the earth. The "New Age" is happening right now and particularly here at Findhorn, for this is where a new "power point" or "light center" has been established by God's wisdom. It is a new center of light, of energy, of the etheric forces which will bring about the changes leading to peace on Earth and a community of human beings whose God, angels, earth and breath are as one.

While I was staying at Findhorn I had established a pattern of leaving my caravan in the afternoon and running to the beach for a short swim. On one particularly sunny and warm day, I took off over the sand dunes and around the gorse bushes and reached the beach at about four o'clock. I took off my shorts and shirt and jumped into the ocean. After coming out, I sat

Continued →

meeting, they told of their experience, but the other car had observed nothing.

The officers were quite definite and certain of what they had observed, with no discrepancies apparent. They were affected by the experience, but not to an untoward degree. I had no reason for doubt.

URI AND THE POWER OF UFO'S by Ira Einhorn

(Ira is a friend and co-worker of Dr. Puharich and Arthur Young, and is known to me personally).

Once again, we are in the midst of another flap over unidentified flying objects. Since late August, every day has been filled with sightings; since 1947, these flaps have emerged at periodic intervals. Anyone who takes the trouble to investigate the data found in such books as Edward J. Ruples **The Report on Unidentified Flying Objects** or Frank Edwards' **Flying Saucers, Serious Business**, to name two of the many reliable books on the subject, will be quickly convinced of two important facts:

Our government has once again lied to us; we are constantly being monitored by extraterrestial civilizations.

Those who wish to track this monitoring back into the dim recesses of recorded history would reward themselves by reading **Hamlet's Mill** by Santillana and von Deschend, which demonstrates the knowledge of a tradition that can't be explained by those who refuse to consider the possibility that man has received help from sources external to our small blue planet. Bruno* was burned at the stake in 1600 for adamantly adhering to the belief in such a tradition.

Consult your Bible with new eyes. Reconsider the talk of the Vimanas in the Vedic tradition. Wonder about the prophecies concerning the New Age, when esoteric knowledge will once again be revealed. That time is now and the UFO's that fill our skies are a physical herald of the possibilities that await us.

(Parenthetically, Ira says that Francis Crick, one of the discoverers of DNA, has just come out with an article in **Icarus** suggesting that man has been "seeded" on the earth from another planet.)

Two other aspects of the UFO situation are of extreme importance to those concerned with the changes we are presently undergoing. Having spent much time with Uri Geller, and having seen him produce evidence again and again that his powers and genuine, I can only says that it behooves us to accept his explanation for his powers. He is a medium for an extraterrestrial civilization that has been monitoring earth for thousands of years. **URI**, by Andrija Puharich, is the story of the development of those powers, and will be utterly convincing to anyone who is capable of reading with an open heart. I lived with Andrija while he was writing the book and was in constant dialogue with him about Uri's powers during that time. Still, the book blew my mind. It is one of the most unusual stories in human history. My editor at Doubleday, who is a close friend, and the agent who is working with the book, were so overwhelmed by **Uri** that they wondered if I would be connected with a fraud. The publishers at the Frankfurt Book Fair (Germany) in October, 1973, had no such problems. Less than two weeks after Doubleday signed up the book, a half million dollars' worth of European rights had been gobbled up.

Stop wondering over Uri and Stanford Research Institute. SRI is frightened and thus still sitting numbly on the biggest story of the epoch.

The second aspect of the UFO situation is their propulsion system, electro-gravidics, Rho-Sigma (a pseudonym) has published a book in German (**Forshung in Fesseln**). Ventla Verlag, Wiesbaden, links UFO's with research done by Thomas Townsend Brown, Dr. Erwin J. Saxl, Dr. Paul Biefeld, Dr. Horace C. Dudley, and Dr. William J. Hooper. I am trying to find a publisher for the book and I am not at liberty to share the information contained therein, in order to protect the author.

(Saxl was a student of Einstein who

---

*Giordano Bruno was a 16th century philosopher who insisted that the universe was infinite in time and space, and that many other populated worlds were in existence. He was burned at the stake for his trouble; his work is still relatively unknown and difficult to acquire.

---

on this deserted beach for a few minutes and then finally got dressed. I began to run back as was my habit, but something stopped me and said "stop and walk to the right." This was away from where I was heading but I did so. I walked into a depression that was surrounded on all sides by sand and as I looked around I noticed a great stillness. The grass was luminously green and the mosses seemed to be like green embers scattered across the sandy floor. There was a fantastic play of light on the sand where the rays of the waning sun were hitting it—like crystals scattering the sun in a sparkling white diffusion of lines. And then I had a very quick realization. It was not a "revelation" but all the experiences and information I had accumulated at Findhorn combined to form an intense image. I saw Findhorn on the planet Earth as a light which had other lights dotted around it. The bluish orb twinkling with lights changed into a body and the light became the Chakras and points on the meridians. And in a moment it changed again with the chakras dissolving to planets and the points to stars and distant galaxies. I saw the light of Findhorn as the brightest light of all, that became the forehead of a man finally diffusing its light throughout the universe and not having any one point as its origin. All during the time of my stay at Findhorn, I was trying to understand what Findhorn was, what was its secret, what was the reason for its almost miraculous ability to transform all that it touched. And then at once, standing in the sand dunes, I could understand the cosmic and earthly spiritual significance of Findhorn as one.

Findhorn is an advance training center for people to learn to handle these new energies. Located away from crowds and pollution, it is ideally situated to assist in the purification of man's consciousness. And not without surprise, you will find these people to be the sweetest and gentlest souls you'll ever encounter. They are living here in a Heaven on Earth, animated by angels and spirits and deities. And they are led by an inner light which shines through the thickest of winter clothing. The destiny and vision of Findhorn unfolds before and through the members in perfect clarity and unity. Here there is preparation, a self-preparation which is simplicity itself in form. It is the unfoldment of human consciousness to join with and receive a transforming light and truth which is concentrated here; a light so real that even the grumbling skeptic is surprised. It is evolving from a light through a center of light to a university of light which will spread across the peninsula of Findhorn and act as a hub to a spoked network of light centers around the globe.

A community exists today, guided by divine truth, manifestly God-like, in care of two perfectly ordinary-appearing people in their fifties. It is peopled by young and old from countries near and far and they display an adaptabilty and flexibility so free that there is almost perfect harmony amongst all of them. If you come to Findhorn, then you had best tuck your guru, mantra, or diet, in your back pocket or leave it home altogether. Instead, bring with you a sense of openness and joy, the same openness and joy which you find here. One might criticize Findhorn over some

Continued →

did experiments on his own demonstrating the relationship between electricity and gravity, and that the relationship does produce a measurable effect. He did it with a pendulum in a vacuum. The others have U.S. patents in this area and the patent numbers are: Brown; 3022430, 3187206, 2949550, 3018394; Dudley has patent 3095167; Hooper has 3610971 and 3656013. Copies can be had from the U.S. Patent office for $1, first class, or 50 cents, second class.)

Electro-gravidics is connected with Einstein's unified field theory, and would provide us with a totally new and safe source of unlimited energy.

ANDRIJA PUHARICH IN INTERVIEW—A FIFTH FORCE IN NATURE: In the October 1973 issue, starting on page 6, **Psychic** magazine has published a long, detailed interview with Andrija Puharich, and also a page of biographical material. Send $1 to Psychic magazine, 680 Beach Street, San Francisco, California 94109 for that issue, which is well worth reading.

OTHER INFORMATION ON UFO'S: For those who wish to learn all there is in print about flying saucers, I can suggest writing to AFSCA, P.O. Box 84, Northridge, California 91324. I am not well acquainted with their information yet, so cannot guarantee its scientific accuracy.

DIRECTORY OF ALTERNATIVE SCHOOLS (published by New School Exchange)—reviewed by Tom Thompson.

In one of the most important areas for changing society—the minds of those people who, because of their age, are forced to attend school (prison)—it is good to see people helping to change the concept of schools from conditioning institutions to places of learning and living.

The New Schools Exchange publishes an excellent bi-weekly newsletter and a yearly directory listing most of the alternative schools in the U.S. and Canada.

A beautiful statement of the Shawnee chief Tecumseh is quoted in the most recent directory: "Sell land? Why not sell the air, the clouds, and the great sea?"

NSE says alternative schools are places where people can go be, learn, interact, dream—rather than public schools.

New Schools Exchange itself seems to be an open, alternative place, inviting change and help from others, and its helpers, judging by the newsletter, seem to be dedicated to freeing people from the traps of institutionalized learning.

Their address is: Pettigrew, Arkansas 72752. Subscriptions are $10 per year, which includes one free ad.

OREGON COMMUNAL GROUP LIVES together and holds public workshops and publishes in new fields of awareness. The group writes: We are the Universing Center, five persons living together in agreement on the methods we use to get through life with more love each new moment we live creatively. The techniques are primal, Gestalt, and Adlerian therapy, tantric and hatha yoga, massage, rolfing, meditation, and we have four retreats this year using these methods. You can write to them for their publications and information at Universing Center Co-op, 3718 S. E. Main Street, Portland, Oregon 97214.

Some of their publications also deal with finding ways to live more easily on less. Sounds like an active group.

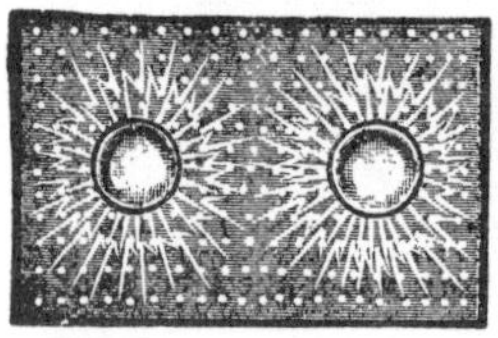

NEGATIVE ION GENERATORS AND THE MULTI-WAVE OSCILLATOR: Paul Hampton, Suite 2709, 88 W. Schiller, Chicago, Illinois 60610, has written to a number of people to report that he is doing research on two electrical devices which have for years been claimed to produce beneficial effects in cases of serious illness. The FDA has outlawed the sale of the devices when health claims are made. Mr. Hampton is working to prove that these devices are indeed safe and effective, and is asking any who have information about them to contact him, and he also offers to report on his research to interested people. A donation or stamps and a mailing label would probably help him, but is not actually required in his letter.

CALIFORNIA SOCIETY FOR PSYCHICAL STUDY, INC. - published a bulletin called **Iridis**. To receive the bulletin, send them $2 at P.O. Box 844, Berkeley, California 94701. If you live near that area, you will be interested in their meetings which are held regularly on the fourth Monday of each month. They also sponsor classes in psychic development by some very gifted psychics. Volume XI, No. 10 of **Iridis** reviews a talk by Alan Vaughn on types of psychic sensitivity.

small aspects but it would be like criticizing a sapling for not being a tree. Findhorn has within it the spirit, form, and nourishment to grow and grow and grow. And it will do so unquestionably. A man without a dream or vision is a spiritual pauper and a planet without a dream or vision is a dead solar satellite spinning through space. Findhorn, tucked away in the north of the world does not have a dream or vision—it *is* the dream and vision of our planet unfolding in life.

IF YOU WANT TO VISIT FINDHORN

Findhorn likes visitors and guests and has accommodations for a limited number. If you would like to visit Findhorn, you should write to the following address:

Findhorn Foundation
The Park, Findhorn Bay
Forres, Moray, Scotland

You may also request from them brochures detailing the literature and tapes which are available. All prospective visitors should write well in advance and obtain confirmation for lodging, whether it be in the Park or during busy seasons in the nearby town. It is fairly easy to get to and they will gladly send to you free of charge a pamphlet called "Introduction to Findhorn" which tells you about transportation to and back, room charges, and a schedule of life in the community. But by all means do yourself and the community a favor by not just "showing up" on the spur of the moment. If you do, you will certainly be disappointed if there is not room for you. If however, your intention is to visit for only one day, then you can come at any time since regular conducted tours and talks are scheduled every day for one-day visitors. Bring your work clothes because everyone is asked to work at something everyday. God bless you

HEALTH & LIGHT-—SURVIVAL AT ISSUE: I have previously mentioned the work of John Ott, director of the Environmental Health & Light Research Institute. Dr. Ott, a pioneer in time-lapse photography (the opposite of "slow motion"), had to make a movie of the development of an apple from bud to ripe fruit. To do this, he used glass to keep the wind from moving the apple, and found that it never ripened!

This led to the important discovery that natural sunlight, which contains medium and long ultraviolet rays, is essential to normal development of many plants, animals and people. Glass filters it out, while artificial lights don't usually have it. Human health might be considerably improved when we have proper light.

An even more shocking discovery is a loss of natural sunlight due to air pollution.

Dr. Ott reports that in Washington, D.C. (not an industrial area), pollution has caused a loss of 14% of our natural sunlight in the last 60 years, and even at such a clean area as Mount Wilson observatory, we have 26% less ultraviolet light from the sun than we had 50 years ago!

In New Jersey recently, squash growers were having trouble with aphids and virus, even in spite of increased use of chemicals. That is, until scientists from the U.S. Department of Agriculture and Rutgers University discovered that laying aluminum foil under the plants overcame the diseases and increased the yield fivefold. Aluminum is a good reflector of light, and an especially good reflector of ultraviolet.

In the November 4 issue of the **National Enquirer** there is an article where Dr. Hugo Graumann of the USDA reports a critical race against time to collect seeds of the wild food plants from which most of our current varieties originally came. He tells that in the last years, many of the original plant varieties that common crops such as corn and wheat were bred from have disappeared, and in the event of diseases to which present varieties are non-resistant, no new varieties can be bred from the original. He assumes "civilization" to be the cause of the disappearance, but it may be that our scientists are missing the most obvious thing in the world—LIGHT!

John Ott's new book, which was to be called **Exploring the Spectrum** is now out under the title **Health and Light**. It is published by Devin-Adair Co., Old Greenwich, Connecticut. Price: $7.50. The book tells about far more than the few items I was able to mention. Read it!

PROFESSOR DOUGLAS DEAN'S REPORT ON PSYCHOTRONICS CONFERENCE - PRAGUE: There was a very important conference on psychotronics and parapsychology in Prague in the summer of 1973. Now the Academy of Parapsychology and Medicine has a cassette tape of a report on that conference by Douglas Dean of the Newark College of Engineering. It can be obtained from APM at 314 2nd Street, Los Altos, California 94022. A $5 donation should be about right.

In his report, Dean tells of many fascinating new developments in the field of parapsychology and physics especially. He tells of Harold Puttoff's movie showing Uri Geller's amazing performances at SRI (on which the official report is still withheld).

FROM SEANCE TO SCIENCE
by George W. Meek and Bertha Harris (Regency Press, London and New York, 43 Oxford Street, London, WCIA IBH, England. Forward by Sir Kelvin Spencer, 1973).

This is a book for the scientist who really wants to know about paranormal events as they actually occur in the presence of one of the best known research mediums presently among us (England). The first part of the book (by Meek) discusses in a fairly simple, yet scientific manner the various paranormal phenomena which have been observed in the presence of such people as Bertha Harris, and which have, by scientific methods, shown to exist.

There are many quotes from well-known scientists, and many matters are discussed which are still quite a puzzle to us such as thought photography (a la Ted Serios), and the Kirlian methods of photographing auras (or energy fields of kinds which have not yet been fully defined).

George Meek is one of the group known as Life Energies Research, which consists of a number of businessmen and scientists with serious interest in a scientific study of the spiritual. The group's interest is sufficient that they are willing to fund people like Meek and Andrija Puharich (I believe—I hope he doesn't mind this reference) in their work of locating the best psychics for research purposes and then learning about their abilities in a scientific way. His book is acclaimed to be Volume I of a series on "Life's Energy Fields."

The second part of the book is written by Bertha Harris herself, and in it she tells in a simple way about her childhood and life as a psychic, giving verifiable details of some of her experiences.

I think the dedication of the book is worth quoting:

"To all of those individuals whose demonstrations of paranormal gifts over the centuries doomed them to suffer persecution and even death because of the unenlightenment of their contemporaries."

The book can be obtained for $6 from George W. Meek and Association, 1412 Jackson Street, Fort Meyers, Florida 33901.

Cash or check and an address label included with your order will help speed delivery.

CHELATION THERAPY HERE: Chelation therapy has finally come to this part of the country. It is a safe method of removing arterial deposits, such as cholesterol, which are the cause of hardening of the arteries, stroke and heart attack. As we presently understand it, people in their forties and beyond have enough deposits in their circulatory system in most cases to produce a risk of the acute problems mentioned, and that risk can be eliminated almost entirely and for years to come by the chelation therapy.

Chelation removes the calcium compounds which settle in the blood vessels, and when that is accomplished, normal healthy circulation is re-established.

The Clymer Health Clinic in Quakertown, Pennsylvania, now has an experienced medical doctor who is administering this therapy. I have been there myself and have had the first of the series of ten chelations, which is the normal course of treatment. Usually a treatment is given about once a week for ten weeks, but that spacing can be varied as may suit the time schedule of each patient.

For financial and other reasons, I expect to wait a month before my next chelation, but it is possible to give the series in as little as ten days if necessary.

Already after only one treatment on May 11, 1974, I have experienced an immediate and lasting benefit in that a varicose vein problem that I have had for over ten years (and for which no reasonable cure seemed to be available) has been considerably reduced. I am expecting that the full treatment will remove the problem entirely, but will give a factual report after I have had that full treatment.

I am told that my father, who is over 80 and has had a heart problem for a long time, could be chelated quite safely in a two-week period, and that five chelations should be sufficient to eliminate the danger of his having any heart attacks in the future, and he should thereafter be able to resume

normal activities of all kinds. Almost anyone whose kidney function is O.K. could safely undergo this treatment and be much healthier as a result.

WIN magazine (**WIN** stands for Workshop in Nonviolence) has always been a peace movement paper. Now for the first time it seems that they may also publish something in the field of the new sciences. I have been asked to write a review of **Beyond Telepathy** by Puharich for that paper. I have done so, and it seems that they will print it. Good that these matters are not left entirely to the military. **WIN** subscriptions are $7 a year, and their address is **WIN** magainze, P.O. Box 547, Rifron, New York 12471.

We have offered plenty of evidence that successful pollution-free cars have been developed, and that little is being done to develop non-polluting energy from the sun. The following article appeared in the Philadelphia Inquirer, July 14, 1973:

SUNLIGHT TO POWER HOME: Energy Crisis is Prime Target by Dominic Sama

Newark, Delaware—Scientists at the University of Delaware are going to use a new house to tackle the energy crisis next week.

The house—heated, cooled and electrically operated by sunlight—will be set up on S. Chapel Street, on the university campus by the Institute for Energy Conversion.

"The house is designed to convert sunlight into electricity and heat," said John Trela, assistant to the institute's director, Dr. Karl B. Boer.

"Approximately 60% to 80% of the total energy needs of the house will be supplied by the solar energy."

Basically, the sunlight comes down and strikes 24 solar collectors placed on the roof. The collectors are wooden boxes built like a duct in a heating system.

"On the surface of the collectors are cadmium sulphide solar cells that absorb the sunlight and then generate direct current (DC) electricity," Trela explained.

"The electricity flows to a battery charger in the basement and it charges a series of lead-acid batteries similar to those in automobiles."

Trela said the electricity from the batteries provides the current for fixed loads like lights, appliances (range and clothes dryer). He said electricity from the batteries can be converted to alternating current (AC) to operate other conventional appliances, like the refrigerator and television.

The frame experimental house, known as Solar One, boasts two bedrooms, living room, dining room and kitchen on the first floor. Two additional bedrooms can be added on the second floor, Trela said.

The house cost $130,000, Trela said, because of its special design and fabrications. No one will "live" in the house except for those conducting the experiment.

"But in approximately three to five years," Trela said, "when the system is mass manufactured, we will be able to build a house at 10% more than the price of a conventional house."

However, Trela added a convincing selling point: "It's clean and there is no pollution involved."

And as Dr. Boer, who is in Paris attending an international convention on solar power, has said, "The cost of sun power will always remain the same."

About 25 other houses have been constructed in the United States over the last decade using solar power. Solar One later will be used as a laboratory, Trela said.

NEW ENERGY SOURCES: John Prenis informs us of a publication called **Alternative Sources of Energy**.

This is a bi-monthly magazine dealing with do-it-yourself solutions to the energy problem. Each issue is a free-wheeling collection of articles, ideas, letters, and fantasies on solar energy, wind power, methane generation, and much more. A year's subscription is $5. Write: **Alternative Sources of Energy**, Route 2, Box 90A Milaca, Minnesota 56353 (Tel. 612-983-6892).

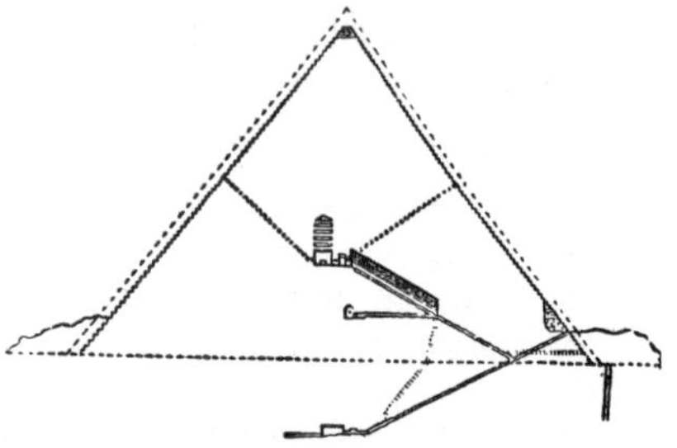

NEW PYRAMID RESEARCH: A new book entitled **Pyramid Power** was recently received from the author, Pat Flanagan, Ph.D. Pat is a physicist, and he has done some very careful work researching the properties of the pyramid form and setting forth new physical theories to explain what he has discovered.

He sent us the book himself, and later we spoke with him at length by phone. Pat offered to send us the book and other materials in order for us to repeat some of his experiments here, and when we have done that, we hope to write of our results in a report (which may be added as a new chapter to his book).

This new book is extremely interesting, and the first thing we have seen on the subject which is really objective and scientific. The book also contains more interesting experiments than any other we have seen. We suggest that those who are interested should write for a copy of the book. You can get it by sending $6.95 to Pyramid Publishers, 438 W. Cypress Street, Glendale, California 91204.

ROSE JONES— follow-up: I have been asked about the case of Rose Jones, who had gone to Tijuana for help in advanced stages of cancer, but who did not go to Hospital del Mar. Rose improved considerably on Laetrile and while taking it, she made radical changes in diet, using large amounts of vegetable juices. Now, 2 years later, she is no longer considered a 'terminal' case.

MIND CONTROL RESEARCH: The Silva organization has recently published two scientific research reports on results of teaching

mind control to special groups. One is entitled "Hallahan High Pre and Post Testing," and the other is called "The Ottawa County Project." Both can be had without charge from Silva Mind Control, P.O. Box 1149, Laredo, Texas 78040.

The first report gives the evaluation of the students in a Catholic girls' high school in which mind control was taught to all of 2000 students.

The Ottawa County Project was one in which it was taught to a group of 41 welfare mothers, all on scholarship, and the results evaluated.

NEW YORK UNIVERSITY—PARAPSYCHOLOGY: Courses in parapsychology are now being offered by New York University.

It is a surprise to us that such an old established institution is giving courses in these new fields, so there must have been a great demand. There are such titles as:

Focusing on new Energies

The Psi Factor: Perspectives in Paranormal Experience

The ABC of ESP

Psychokinesis: Mind Into Matter

Psychical Research: Myth Into Science

Altered States of Consciousness

After Death, what?

Implications of the Human Psi Factor

Some of these appear to be one-day events, and some are longer courses, all apparently open to the public, some with charge, and some without. For information, call (212) 598-2373.

CANCER AND LAETRILE: Once again, there is a news item about Laetrile. The Sloane Kettering Cancer Research Institute has finally begun to test its effects scientifically. The details are published in the December 7, 1973 issue of **Science**, which is the paper of the professional scientist in America.

The Green Revolution Volume 12, No. I, is entitled, "Perspectives on Major Problems of Living." In it, Ralph Borsodi writers:

"The Green Revolution is a social, economic, and political movement. Unlike most movements, it does not begin with clamoring for new laws: it begins not with political agitation, but with persuading people to themselves live in accordance with a culture fundamentally different from that which is now prevalent. . ." Available for $1 from the School of Living, 442 1/2 Landfair Ave., Los Angeles, California 90024.

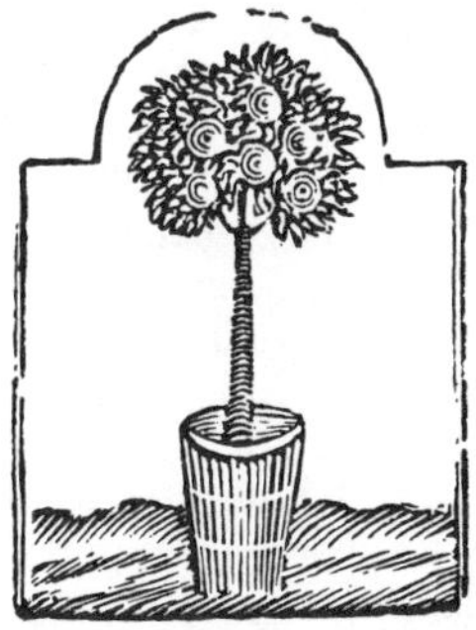

The FIRST FRUITS COMMUNITY is, we gather, looking for fruitarians who wish to move (eventually) to a tropical mountain climate to enable them to live on fruits and (they believe) learn to see more deeply into the mind of things. Send a self-addressed envelope to Nadine Forrest MacDonald, 8300 Abbott Avenue, Miami Beach, Florida 33131 for a newsletter.

THE LITTLE FREE PRESS is a sheet of inspiring commentary and cartoons about peaking out into alternatives: "As one drop of water expands into vapor and fills a whole room, could a good idea expand and be passed onto all the peoples?" 2¢(!) each or $1 for 8 issues. Little Free Press, 71 S. East 14th Street, Minneapolis, Minnesota 55404.

SERVAS INTERNATIONAL: Open doors around the world. One of the most exciting experiences a traveler can have is to live with a family in a foreign country. For those who cannot travel, the next best thing is to bring the world into your living room by inviting a foreign traveler to stay with you. SERVAS, a non-profit, non-political and interracial system of hosts and travelers around the world, can help you do this. "Similar to the Traveler's Directory, SERVAS has hosts in 35 countries, including, of course, the United States. The difference is that one does not have to be a host to be accepted as a traveller. You do have to be interviewed personally or submit a resume. They have at least eight interviewers in Canada. The organization requests a $25 donation from people using their host listings. Write and send self-addressed envelope to Lesma Hossack, Box 6114, Station A, Toronto, Canada, M5W 1P5 or U.S. Servas Committee, Box 790, Old Chelsea P.O., New York City, New York 10011.

SPECTRUM: Journal of the occult: A new concept in total reporting! An occult/UFO/Fortean information outlet featuring factual news events in all these fields, scientific and analytical articles by well-known authors, national and international $4 yearly on a quarterly basis—foreign $5. Send subscriptions to: Barbara Lucas, 6290 34th Avenue, North, St. Petersburg, Florida 33710. Send 25¢ for sample. Edited by Joan O'Connell.

SCHOOL OF THE ARTS, a small liberal coed high school on beautiful New Jersey farm; can accept students with need for individualized program in all subjects. $3,500 per year total fee. Some scholarships. More male students are needed especially. I visited there, and was impressed with the free and open-ended approach to creative learning. During the current year or two, the School of the Arts (which is more scientific than artistic) will be held in a commune in Philadelphia. Write: School of the Arts, Box 114, Stillwater, New Jersey 08075.

PYRAMID EXPERIMENTS: Pat Flanagan, author of **Pyramid Power**, has been very prompt and generous in supplying us with all the materials necessary to follow through on experiments with pyramid shapes. However, when we asked for the exact details on the experiments so that we could repeat them in just the way he has done, he never wrote or returned our call so that we cannot verify any of the information. We did try the experiments with sprouts and found no appreciable difference in rates of growth between a regular control batch or of a "pyramid energy" device. Until we get more definite advice and are able to verify the results, we suggest not bothering about the book **Pyramid Power**.

"GOOD NEWS HOTLINE provides good news for people who are sick and tired of being sick and tired." Reports some interesting items such as:

In Los Angeles, TV, radio, and papers had a field day when Marilyn Monroe took her life. Within a month, the suicide rate went up 40% But in Boston during a newspaper strike, the suicide rate went down 22% and in Detroit, with a 268 day newspaper blackout, the suicide rate dropped 28%. If bad news is bad for you, shouldn't good news be good for you? They also publish "How to get what the government owes you": 100 pages of listings of free and almost free stuff from the U.S. government. The book and a nine month subscription to **Good News** can be had from Good News Hotline, 1818 Whipple Avenue, N.W. Canton, Ohio 44708.

ISRAELI DOCTORS STRIKE? Hearing about the drop in suicide rate when papers strike reminds me of a news item we heard some time back. Report was that when medical doctors in Israel went on strike for several months, the death rate in Jerusalem **dropped** by 50%. We have not confirmed this report, and wonder if it is true.

NEW RESEARCH: **The Fields of Life** by Dr. Harold Saxton Burr, late of Yale University, has recently been published by Ballantine Books as a paperback book ($1.50). In it, Dr. Burr described measurements of the electric voltages that surround the human body and other living things.

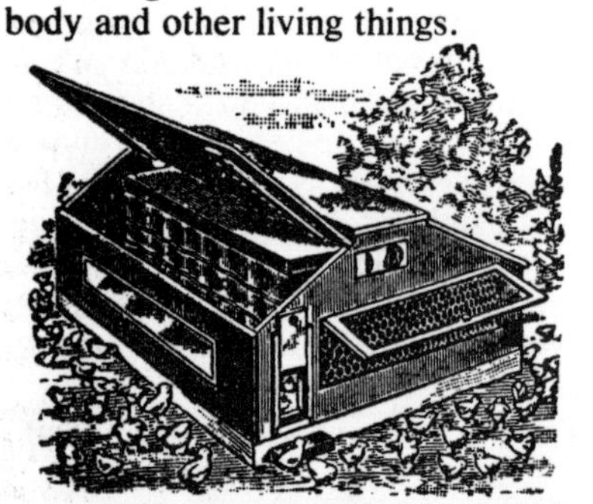

OVULATION: THE FEMALE FIELD: In his study of human fields, Dr. Burr discovered that there was a distinct difference between the fields of men and women. Measuring the voltage between a person's index fingers, he found that in men it remains relatively constant, but in women, there is a sharp rise lasting about 25 hours each month. Considerable research into this phenomena confirmed the fact that this rise was connected with the time of ovulation. If a woman did not ovulate, the rise did not occur.

A most interesting observation was that just about 30% of the women tested were found to ovulate within just that time which the rhythm method of birth control would indicate as the fertile time. A much larger percentage of women were found to ovulate at other times which did not seem directly related to those calculations, as suggested by Drs. Knaus and Ogino.

Burr and Ravitz (one of his coworkers with whom I spoke by long distance telephone) were not aware of the cosmic fertility times discovered by Jonas.

Dr. Jonas, who developed astrological birth control, found that women who abstained during that time had only a 15% chance of getting pregnant in a year.

It is also generally accepted that only about 30% of women are successful in using the rhythm method alone for birth control.

Could it be, therefore, that some women (about 30%) do ovulate only in the rhythm time, whereas others might ovulate at the cosmic period, and that a way might be developed to detect which group a woman is in, or what other factors are involved in this determination?

ELECTRONIC DETECTION OF OVULATION: In an effort to learn how this might be done, we have built the kind of electronic equipment which Dr. Burr described in **The Fields of Life**. We have been testing the equipment ourselves, and have soon to make a trip to New Haven to visit with a Yale University scientist who had done similar work to learn the best way to use it. When that is clear, we have a promise from the director of Planned Plarenthood that they will ask their staff of 20 women to take daily measurements of their own electrical potentials for a month.

We expect that many months of testing and comparing with other signs of fertility will be necessary to find out if there is a practical way to detect ovulation by such measurements.

Burr's associate, Dr. Ravitz, also suggested that some women may tend to ovulate at certain times when certain organs are stimulated (as is common with rabbits). I have myself wondered if such a theory might explain the effectiveness of the astrological birth control method. Perhaps the current project will give us some answers.

HEALTH AND LIGHT: On our visit to John Ott in Florida, he gave us literature on lenses for glasses that allow all of the sun's natural ultraviolet and, of course, visible light to pass through. He indicates in his book how important that is to health if you have to wear glasses. Normal eye glasses screen out all ultraviolet.

An excellent article on the above subject entitled "A Reevaluation of Ultraviolet as a vital part of the Total Spectrum," is available from Obrig Laboratories, Inc., P.O. Box 1899, Sarasota, Florida 33578. No price is given, so it is probably free.

For the Full Spectrum eye glasses, write, Armorlite Lens Co., Inc., P.O. Box 1038, Burbank, California 91505. For contact lenses of similar quality, write Obrig at address above.

FINDHORN IN AMERICA: Findhorn (in Scotland) is probably one of the foremost New Age communities of over 100 people, endeavoring to live, as well as spread the coming of the new day. They have now sent some of their most experienced members and others to America, and to begin with, they have started a group in California known as the Lorian Association, P.O. Box 941, Belmont, California 94002, (415) 591-4611 and 591-6652.

Though not yet living communally, in California the Lorians are already presenting programs for the public, and it is our hope that they will soon be able to come here and present their programs in East Coast cities and learning places.

ARF VISITS WITH CLEVE BACKSTER: On Thursday, February 21, 1974, Bruce Dickson and I went to New York City to meet my brother Bruce at the laboratory of Cleve Backster, and we were able to observe some of his latest research. We found that instead of using the famed philodendron plant, Cleve was now getting responses from a testube of yogurt (to which he makes contact by immersing pure silver wires in the yogurt). The yogurt had to be at warm temperature (about 90° F.), maintained by a thermos flask in which the test tube was immersed. Responses were picked up on an electroencephalograph, which did not supply any current to the sample, but only recorded the voltages that the yogurt produced.

We saw continual, very small variations which, however, became very large at times, such as when two of us carried on an emotional conversation about commune experiences. We taped the conversation, then played back the tape to note where the response occurred. Major response occurred on two occasions near the start of the 15 minute tape, and I am aware that on at least one of those instances, I had experienced a strong emotional feeling at the exact time of the response. However, for the experiment to be scientifically significant, it would have had to be repeated many times, and we did not have time for that.

We were shown results of a previous similar experiment which Cleve told us had more repeatability. In this, human sperm was used in place of the yogurt, and the human donor of the sperm was given a capsule of odorous amyl nitrite to break, upon which there are generally sharp voltage peaks from his sperm in the test tube. Cleve said that this occurs before the odorous compound is smelled by the person.

An experiment of this type seems simple to reproduce, and we are hoping someone can build or supply D.C. amplifiers and power supply for our chart recorder (dual unit) to help us set up and duplicate the experiment. One trouble Cleve was having was that the silver wire tended to coat with a polarizing compound. We supplied him with some gold wire and some silver chloride electrodes to try instead.

PREVENTIVE MEDICINE: MICHIGAN: A few days ago, we had a very important phone call from a foot surgeon in Flint, Michigan, by the name of Dr. William DeHart. He told of his visit to Dr. Alsleben in Anaheim, California, where he himself underwent chelation therapy for the removal of calcium and cholesterol deposits from the interior of his arteries. He found that this so improved his own health, that he instituted a program of chelation therapy for his own patients. He also told of the use of hydrochloric acid to eliminate bacteria from the bloodstream.

In this therapy, which precedes chelation, just 50 cubic centimeters of saline solution containing at most one part in a thousand of HCL, is given intravenously, with amazing effect of safely destroying vast numbers of bacteria by reactivating white cells. He said that a strep throat would clear up in four minutes.

Dr. DeHart also told of a book entitled **Three Years of HCL Therapy**, which was published in 1935, in which we are told of pharmaceutical interested offering money to researchers to stop work on HCL therapy because it was too cheap to be a profit to them. He is sending us a copy of that book for our study, and we will report further. We want to get help in this from other doctors.

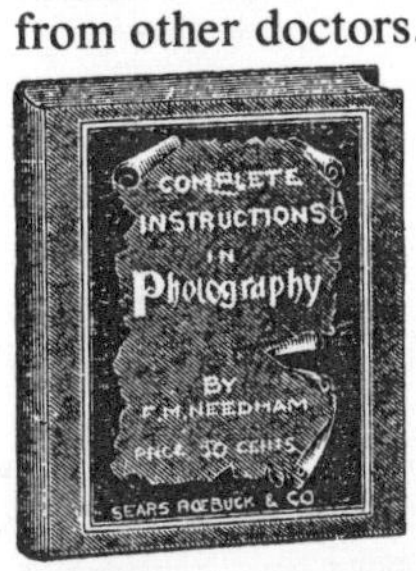

KIRLIAN PHOTOGRAPHY BOOKS: We have received a small 58-page instruction book for the making of photos of "auras" of leaves and fingers and the like, or what is commonly known as "Kirlian Photography." It is the first book we have seen that actually tells you how to do it, and offers all information and equipment required. The booklet, called "Kirlian Electrophotography," is sold by Mankind Research Unlimited, Inc., 1325 1/2 Wisconsin Avenue, N.W., Washington, D.C. 20007. (202) 337-7370. Price $4.75. Our one question regarding this publication and the equipment which MRU offers is, why is it so expensive? We understand that the Tesla coil units, which they suggest using, and which are priced from $40 to $70, can also be had for considerably less elsewhere, but we do not at the moment have that address.

We are expecting that another book by Lynn Schroeder and Sheila Ostrander will be out with all the details on low cost Kirlian equipment. It will be publicized, and we will supply the information when we can.

AMERICAN SOCIETY FOR PSYCHICAL RESEARCH: in New York City is one of the most scientific of the research groups in new studies of psychical science. I spoke with Dr. Karliss Ossis, ASPR's director, a few days ago about their most recent work, and learned that in their study of "out of body travel," using subjects who are able to perform that at will, they are setting up optical experiments to determine not only if the sleeping person can leave his body to view objects in another room, but also from what angle he views them. Additionally, a physicist on the staff has been working with field effects to see if the presence of the "astral body" can be detected with any kind of electrical field effect equipment.

ALPHA-THETA BRAIN WAVE MACHINES: of highest quality and low cost are still available at $50 and $70. We get them direct from the manufacturer at a 35% discount and have them shipped directly to you by air at our expense. Other well-built auxiliary equipment, such as brain wave signal generators, percent of alpha meters, isolating amplifiers, can also be had at similar low cost. Send 25¢ or stamps for brochure to ARF. Ask for brain wave information.

VIDEO TAPE RECORDERS FOR LESS: Bargains in videotape recording equipment can be had from Communications Unlimited, P.O. Box

463, Whitmore Lake, Michigan 48189 (313) 449-4367. A Bell and Howell unit selling for 995 can be had new for $450, or used for $250. These machines are probably not the latest models, but good. Accessories also.

BODY FIELD METERS: will be ready in about two weeks. We are building ten of them, and will make them available for sale or rent at the end of March. The meters operate on a pair of nine volt transistor batteries for a long time and can measure the voltage (in millivolts) between parts of the body. The meter only uses a billionth of an ampere from the body for full scale reading. Price, including postage, will be $60 for the regular units, $70 for units with an extra high sensitivity scale for plant measurements, etc. Rental will be ten dollars for the first month plus $5 or $6 per month additional. A deposit of the full price will be required to rent the meters.

PUPPIES DUE MARCH 31: Our beautiful black German shepherd dog, Laurie, is due to have puppies soon. We expect them to be born about March 31. The mother is very friendly and good-natured.

EDUCATIONAL TV ON PARAPSYCHOLOGY: In the newsletter of the American Society for Psychical Research (5 West 73rd Street, New York, New York 10023 (212) 799-5050), the following item appears:

The 26-program TV series on parapsychology, "Who is Man?," featuring Herbert B. Puryear, director of education of the Association for Research and Enlightenment, has recently been made part of the Public Television Library catalogue used by station managers and program directors. If you'd like these films to be seen in your area, you can contact the manager of your local public television station and request it.

Using a basically conversational format, the programs are designed to give insights into the general field of parapsychological research, and explore the relevance of such research for man's understanding of himself.

SHORT NOTICES—THE DOUKHOBORS: The Doukhobors are a large semicommunal group of people living in Canada who were some of the original pacifist followers of Leo Tolstoy. Few of the public know about them. But now they have been publishing a magazine for their young people which gives a clear insight into Doukhobor life. It is called "MIR" (Russian for peace), and can can be had from MIR Publications, Box 730, Grand Forks, British Columbia, Canada. The price is $4 for five copies, or $7.50 for ten. Published by volunteers six times a year.

NYU TEACHES HEALING: The largest nursing school in the country, New York University School of Nursing, has, according to the enclosed clipping, begun to teach their nurses to heal by the "laying on of hands" and other seemingly "psychic" methods.

AMERICA'S LARGEST NURSING SCHOOL IS NOW TEACHING PSYCHIC HEALING: By Ron Caylor.

Psychic healing is now being taught at America's largest training school of nurses.

The subject was added to the curriculum this year at New York University Division of Nurse Education, and 130 graduate students are studying the healing art of the "laying on of hands."

"There's no question that this healing method works—it works incredibly well," Dr. Martha Rogers, head of the division, told the **Enquirer**. She gave her wholehearted approval to the unusual training after being cured of a serious respiratory ailment by the healer who now teaches the class.

"It has remarkable implications for nursing—and I can forsee the time when we'll extend this training to all of our students," Dr. Rogers said.

Dr. Sidney Borowitz, chancellor of the university, told the **Enquirer** that Dr. Rogers' decision to offer the subject has university backing.

"We have the autmost faith in Dr. Rogers—she's been head of the division since 1954 and is considered extremely capable," he said.

The healing art is being taught by Dr. Dolores Krieger, 50, an associate professor of education who has been practicing laying on of hands for about two years—with great success.

"I think laying on of hands will soon be accepted everywhere as a normal part of hospital routine," Dr. Krieger said.

"There's no limit to what it can cure—arthritis, fever, backache, anything you care to name."

"It's something that can produce results when all else has failed."

The graying teacher, a registered nurse but not a physician, emphasized that laying on of hands is not "faith healing."

"Faith is not important in laying on of hands—it is a natural physical process," she said. "And there is no abracadabra or hypnotism involved. But we have just begun to understand to what extent it can be used."

Dr. Krieger's students now receive two hours' instruction a week during a general course on nursing practice, but plans are being drawn up for a full-time course to begin this fall.

Students at the school, which is affiliated with University Hospital, New York, are assured by Dr. Krieger that anyone can successfully perform laying on of hands.

"The two main essentials are practice and experience. In addition, the healer must be a fairly healthy person," she said.

"The diagnosis itself is the most difficult thing to teach—it's something that comes only with experience. I've found that in many cases the problem doesn't lie in the area where the patient feels pain.

"I tell my students to use their hands like divining rods and to try to feel some sort of sensation.

"Some people feel pricking sensations in their hands when they pass them over an area of a patient's body where a problem lies. As for myself, I feel extreme heat in my palms. . . sometimes I also feel something like a magnetic pull."

"Once I've found the area where the trouble lies, I lay my hands on the patient for between 5 and 20 minutes. I think consciously all the time about transferring energy from my body to the patient's."

Dr. Krieger said she first became interested in laying on of hands while working with noted healer Oskar Estebany at Pumpkin Hollow Farm—a private research foundation in Massachusetts—in 1971.

"I saw some healings take place which couldn't be explained away by the medical training I'd received," she said. "I was puzzled."

"I began studying Eastern literature, particularly that of India and Tibet, and found that laying on of hands reaches back into antiquity.

"I learned there's a force that's called 'prana'—which we would call vigor within all of us. Healthy people have more prana then they need, while ill people don't have enough. I discovered prana can be transmitted from one person to another by touch and the desire to heal—and that, incredibly enough, anyone can do it."

Dr. Krieger's powers of healing were attested to by Dr. Rogers and two other university staff members.

"I had been utterly miserable for weeks with a bad respiratory infection when I asked Dr. Krieger for help," said Dr. Rogers. "When she touched me, it cleared up and I felt magnificent."

Dr. Leah Gorman, an associate professor of education at NYU, told the **Enquirer**: "I had strained a ligament, and my back ached constantly. When Dr. Krieger placed her hands on me, I felt heat—then the pain disappeared altogether. And it has not returned."

Rose Marsoobian, the school's office supervisor, said she had suffered from cramps: "After Dr. Krieger touched me, I felt like a new person."

The healing technique has drawn support from others outside the university. Dr. Michael Schacheer, director of the Rockland County (N.Y.) Community Mental Health Center, told the **Enquirer** Dr. Krieger's successes are "very exciting."

He added: "Laying on of hands." should not be dismissed lightly—it's a scientific process that really works."

HARNESSING THE WIND FOR ELECTRICITY: A network of electricity-producing windmills along the New Jersey Garden State Parkway that could generate as much or more power as the proposed offshore nuclear power plant could be a reality without any threat of radioactive pollution.

Rosemarie Totaro, New Jersey assemblywoman, has introduced this proposal in the New Jersey legislature.

With the expert assistance of professor William Hieronymus, University of Massachusetts, Amherst, the Parkway Windmill plan was developed. Professor Hieronymus has been involved in energy research for many years and has developed sound engineering studies under National Science Foundation grants that could

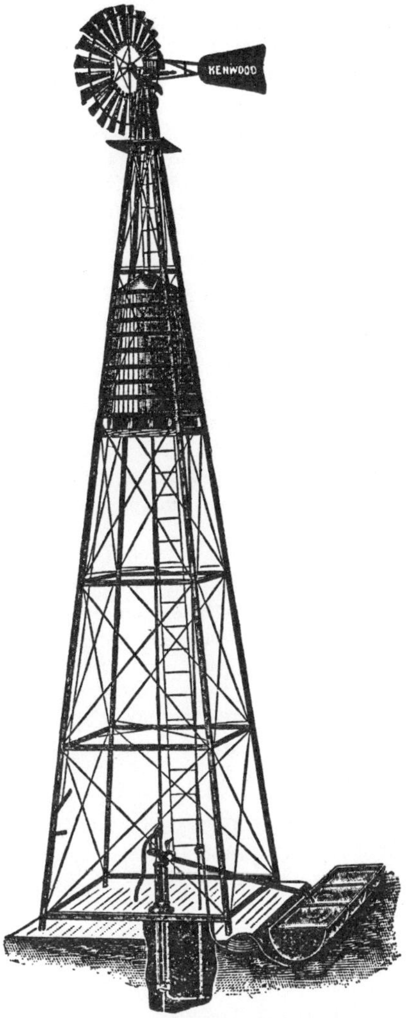

provide all this nation's electrical needs for renewable energy sources, solar and wind.

Hippocrates said, "For they praise what is outlandish before they know whether it is good, rather than the customary which they already know to be good; the bizarre, rather than the obvious."

The question is: Can we break away from the so-called "conventional" method of generating electricity by using steam produced by burning a diminishing, non-renewable fuel, be it coal, oil, gas or uranium, and embark upon the only course that insures the survival of the human race?

That is the question. There can be only one answer—we must.

John K. Mustard.

For additional information on this subject contact: Task Force against Nuclear Pollution, 305 High Street, Moorestown, New Jersey 08057.

GROWING FOOD YOURSELF: It is possible to improve nutrition and health and save a lot of money, possibly without much expenditure of time. Florence Malin, whose health has been remarkably restored by the work of José Silva, combined with that of Ann Wigmore, visited us and told some of her experience, so I asked her to write a bit about it to send out to you: The following is her account:

HOW I EAT NUTRITIOUSLY ON THREE DOLLARS A WEEK:

The top two stories of my life happened in the last two years. One: taking the Silva Mind Control course (cost $175). Two: getting detoxified and rejuvenated at Ann Wigmore's Hippocrates Health Institute in Boston for two weeks (Cost: $90 plus bus fare). Always having known that "there is really nothing new in this world," at age 54 I am still amazed that I had never encountered these two unique experiences in the past.

To the Silva Mind Control course I traipsed—because of tremendous difficulty with concentration and memory—a result of the scars of living so long. The concentration improved tremendously, the memory more slowly. But the big surprise bonus was the disappearance the first week of my progressive, painful arthiritis—except for the tender spot of the bottom of my spine.

Then, to the Hippocrates Health Institute last August I ventured to (1) analyze my stomach complications and (2) find out how to eat nutritious meals of sprouts, greens, and wheat grass on three dollars a week. In my tote bag I carried my old faithful mind control—relaxation tapes to assure me a vigorful vacation. Well, that must have been a potent formula because not only did I get rid of my gas pains—but in two weeks I got perfect vision when for 40 years I had been nearsighted! In addition, I lost 13 pounds, the pain at the bottom of my spine, a lot of my loose neck skin, the plaque on my teeth; my complexion glowed like a baby's, and I felt the energy of a bomb. The vision change was particularly astonishing because I had been doing eye exercises and mind control programming for this for over a year. But I got more out of the Boston trip. I now know, having studied the Wigmore nutrition research for seven months, how to live healthily for the next 50 to 70 years (at least). The pressure of aging and preparing for retirement are off! I know that in seven years I shall have a younger body. The pressure of getting cancer again (I had it removed surgically 13 years ago, but the fear had remained) is off.

And, since I am now successfully

growing my own fresh food in my small apartment, the economic pressure of needing a big salary check is gone. I am absolutely convinced that you can feed yourself most nutritiously for under three dollars a week —if you can conquer the food addiction habit and retrain your taste buds which have been programmed since birth. My mind control exercises gave me tremendous relaxation results and mental discipline for general health maintenance, but no one was able to impress upon me specifically the damage that chemically-laden junk food was doing to my cells until I myself reversed the process last August. What a startling revelation! I'm so lucky! And I used to believe that doctors had the ability to restore health!

My bit about cancer: In Boston I actually observed cancer cases in the process of healing. And I knew it was so. Having read much in the last few years about the application of drugs towards cancer cures, I now sincerely believe that live food (such as bean sprouts or wheat grass juice) is the best medicine because it rejuvenates the cells. Drugs provide no nourishment and are generally destructive to cells. All cells know how to function properly if they are nourished properly. And wheat grass is a complete food.

I feel so strongly about body detoxification and cell rejuvenation that I am now preparing to embark on an auburn hair-growing escapade. My hair is now gray. I am preparing 30 programming tapes on nutrition. If I got perfect vision using my two old faithful tapes in Boston, what will happen with 30?

My favorite quotation this year is by Viktor Kulvinskas, a director of the Hippocrates Health Institute and author of **Love Your Body**—"No one knows what he can do till he tries."

REAL BOOK OF FAIRIES: We had a visit from A.C. Paddock, who has been looking into unusual phenomena for many years. This time, aware of our interest in fairies, he brought us **The Real Book of Fairies** by Edward L. Gardner.

This short book contains five photographs of actual fairy creatures. The photos were taken by two young girls in England in 1917 and 1918. The whole affair was investigated by a co-worker of Sir Arthur Conan Doyle, who is well known for the Sherlock Holmes works. Doyle was also a scientist and a researcher of the psychic.

The photos were studied for possible fraud by some of England's best photographic scientists, and although the scientists did not confirm their validity, they could find no evidence of fraud. The last three, taken a year after the first, were also taken on special plates which had been marked in advance by the manufacturer with a mark unknown to the recipient so that no pre-fabricated plates could be substituted by trickery. As the girls reached puberty, their ability to relate the fairy creatures decreased and no further pictures could be gotten.

The evidence from the extensive study which was conducted seems to indicate that creatures exist in a spirit form who are, at times, able to materialize in a visible body; that this could be the basis of the centuries-old legends about fairies which have existed in many lands.

We are still looking for people who are able to see creatures of this order, but we now realize that the testing may not be as easy as we had supposed.

(**The Real Book of Fairies** by Gardner can be obtained from the Theosophical Publishing House in Wheaton, Illinois for $3.25 or from the same company at 68 Great Russell Street, London, ECI.)

CHELATION THERAPY IN PENNSYLVANIA: In the last newsletter, I mentioned chelation therapy with EDTA as a means of overcoming arteriosclerosis, lead poisoning, and the effects of other pollutant poisons in modern living.

Since then, I have discovered that EDTA is the commonest method of removing lead from children who have been affected by it, and this is done at a cost of about $12.00 worth of EDTA, given in five intramuscular injections (with anaesthetic mixed in to prevent pain). The compound is also available in oral tablets for administration to adults. I do not know if these doses or forms are helpful in cases of arteriosclerosis.

However, I have been able to speak briefly by phone with a doctor near Pittsburgh who learned the work form Alsleben, and treats patients with EDTA in various ways (with professional testing), and at a cost similar to that of Dr. Alsleben ($750 to $1500, or $2000 if hospitalization is required). I need the help of a medical doctor or researcher to discover the reason for the discrepancies, and also the value of the treatment.

OTHER NEW THERAPIES STUDIED: At one time chelation therapy in this area was only a hope. Now it is a reality. In a few months, we may also be able to talk of two other therapies that seem to be quite as remarkable as the above or even more so. I am presently studying a book entitled **Three Years of Hydrochloric Acid Therapy** edited by Dr. Henry Pleasants, Jr., A.B., M.D., F.A.C.P., and written by Dr. Burr Ferguson and Dr. Walter B. Guy and others.

This book is out of print, and seemingly unobtainable. It was supplied to me as a Xerox by a doctor in Flint, Michigan, and I have copied it for the Clymer Health Clinic. This book is so important because it gives details of an extremely effective, low cost therapy which produces a remarkable increase in the body's resistance so as to overcome almost any infection and rapidly heal many otherwise dangerous conditions.

This therapy has also been used by Dr. Rudolph Alsleben, M.D., in his work and is very favorably described in his book, **How to Survive the New Health Catastrophes**.

HCL (hydrochloric acid) is normally produced in the stomach and is required by the digestive process. In good health, this acid is present also in the blood and acts to stimulate the white blood cells to fight infections, cancer cells, etc.

According to a long series of articles in **Medical World** in the 1930's, many doctors found that injections of small amounts of diluted HCL could produce a tremendous and immediate increase in the natural defensive activity of the white blood cells— and in that way all sorts of infections and other diseases could be overcome—even cancer. For those wishing to test this, the usual dosage was one part of chemically pure HCL to between 500 and 3,000 parts distilled water. Five to fifty cubic centimeters are injected intramuscularly or very slowly into the vein so that it circulates as mixed with the blood. A rapid rise in number and activity of white blood cells is the usual result, and this brings about increased resitance to disease and rapid healing.

We would like to tell you where you can obtain the book, but that seems impossible until it might be republished.

From all the evidence given, it seems that this HCL therapy is so safe and effective as to be almost a miracle: it also is extremely cheap, and we suspect that is the reason it is not used. Quoting from the book (copyright long expired):

"As I look over these and many more similar records (of healing) in my notebooks a sincerely sad thought comes to me: What a pity that this excellent, 100% dependable therapeutic agent, this HCL solution injection, may never become generally accepted and used due to an inherent, terrible and unpardonable defect; it has and carries an indelible birthmark as greatest of all sins: 'It is just too common for anything! There is no mystery about it; as easily procurable, as easily applicable—accepted or not accepted, listed or not listed—in all parts of the globe, thus: it cannot lend itself to profiteering commercial exploitation and still less to monopolistic purposes. Of course, we mean 'ethical exploitation' and very 'ethical monopoly'."

"Dr. Burr Ferguson and other advocates of HCL solution injections are simply in bad luck, for they could not 'hire out' this substance to a big, powerful, and respectably ethical laboratory concern for its 'exclusive' manufacture, and make one and one half million dollars a year [1931 figures] profit on it, ethically divisible between the laboratory and the doctor discoverers, as it happens with another exclusively 'licensed out' [and of course, 'accepted'] product.

"It also happens, of course, that —the inventors and doctors being, usually, poor businessmen—the laboratory, big and powerful and all that, very ethically gets the lion's share! In short, there is the reason!"

While typing this, I called Dr. Poesnecker at Clymer and he told me that he had read half the book already, and that they will soon be experimenting with HCL therapy at their clinic. He also told me that with regard to the chelation therapy, others had also found that it can overcome the varicose vein problem as well as arteriosclerosis and the like.

As work in these therapies goes forward, we will learn and report, and if necessary, we may publish the information more fully when we have our own evidence to back up what others have written.

WARNING ON HCL THERAPY

HCL should not be used indiscriminately. Infections, colds, etc., should be treated with large doses of vitamin C, garlic, etc, but HCL should only be used as an emergency measure. A detailed pamphlet entitled **Hydrochloric Acid Therapy** will soon be available for one dollar from ARF.

THE FARM OF STEPHAN GASKIN: On our way back from California, we were able to stop for one day at the farm of Stephan Gaskin, dren). The address is: The Farm, Box 197A, Summertown, Tennessee 38483. (615) 379-7248.

We happened to have the book **Monday Night Class** with us on the trip, and I was able to read it while others drove. The book is just a direct transcript of the class, questions and answers on every subject. Unfortunately, we arrived on an "All Farm Day" in which everybody who can is put to work on some urgent farm task. This time it was a green bean harvest. As a result, there was no time to really show us the whole place. The farm people wanted us to stay at least three days, and I would have preferred that, but others of our group felt an urgent need to get on to Philadelphia, and so we had to proceed.

An additional joy was to meet there Luke Stangl, who is about 21 and married and a father. I last knew Luke as Lucas in the community of the Society of Brothers in Paraguay. He was eight years old at the time, but remembered me quite well.

I could write a whole page about my day at the farm, and I was very impressed by their commitment to complete honesty in all matters. I also took a number of photographs, some of which I sent on to Luke's mother, who is at the Society of Brothers community in New York State.

Stephan is a teacher, or "guru," author of **Caravan** and **Monday Night Class**.

He taught for a time in the San Francisco area, and in the end, invited all his students to come and live with him on a farm. They now have about 550 people in all (including about 140 chil-

CANCER-LAETRILE-MEXICO: In Tijuana, Mexico, I visited the Hospital del Mar, where I had been two years before, and where many cancer patients come for treatment with Laetrile. Arriving in the early evening, I located Dr. Madrid, with whom I had talked earlier, and he invited me to a meeting with other doctors and friends. As luck would have it, few of the others showed up, and we had a good opportunity to talk about many things.

Two years ago, Dr. Madrid had spoken very positively of the work there, but also told me he planned to leave it soon. Now, he told me, the work was not going so well; that either the quality of the Laetrile was not what it was, or, for other reasons, there was a failure to control the cancer in many patients. Sometimes, when patients died, relatives were bitter about the amounts of money that they had spent on treatment. He said the hospital was planning to start using some other therapies in addition to Laetrile, even radiation, which is not highly thought of by many in underground medical circles.

Next morning, I went and spoke with the medical director of the hospital and he spoke cautiously. He confirmed some of the things Dr. Madrid had said, putting them in a slightly more positive light. I then told him about the work of Dr. Alsleben and left him some of Dr. Alsleben's literature and suggested that there might be a connection between cancer and the poisons that Alsleben was eliminating from people's bodies. Some of Alsleben's experiments seem to indicate that.

Later, I made an effort to contact Dr. E. T. Krebs, to tell him about the situation in Tijuana, and about Alsleben's work, and found that he knew about the difficulties at Hospital del Mar, and was a friend of Dr. Alsleben's already.

ARTHRITIS—CAN OLD AGE BE CURED? About three years ago, a man came here and told me that he had had arthritis for some years, but that it was now completely cured. He explained further that he had cured it himself with novacaine injections he learned about in the library of the College of Physicians (19 S. 22nd Street, Philadelphia, Pennsylvania). He told me I could find the articles on novacaine therapy by Dr. Anna Aslan by asking for stack number PAM 732.I.

Recently, a woman came to us asking if we knew anything to help in arthritis and so we suggested that she go and read the material and copy it for us.

When this was done, we had a long report of medical research by doctors in Rumania (C.I. Parhow Institute of Geriatrics, Bucharest) and this report indicated that Dr. Parhow and others had started with the theory that "old age" is a treatable disease, and found through long research that injections of novacaine in the muscles, veins, or arteries, depending on circumstances, could bring about hormonal and other changes that seemed often to reverse the aging process. The therapy was dramatic but slow. During a period of several years (three) all sorts of conditions were reversed. Some bald people grew new hair, and it wasn't gray; arthritic conditions and senile symptoms cleared up. There were far less deaths among the old people so treated then others who did not receive this therapy.

Now, as it happens, we have received literature indicating that an oral capsule containing novacaine together with other catalysts and vitamins is on the market in Europe, but is not yet available here except at a rather high price (three times the English price). It costs about $1 per day to obtain these capsules (called CYTO-H3) in this area.

MORE FAIRIES

Kingdom of the Gods by Geoffrey Hodson (Theosophical Publishing House, Box 270, Wheaton, Illinois 60187, $10).

Since the book on fairies by E. L. Gardner, I have read another book with more information about spiritual kingdoms.

The author claims to be a clairvoyant who can see angels and nature spirits and who has traveled around the world observing them, noting their function and appearance. Though no evidence is given, we may accept what we feel comfortable with. My personal feeling about the author is that he is a critical observer aware of the limitations of his vision, however expanded. His powers of observation seem keen, and his attention to detail rigorous and dispassionate.

The 250-page book includes a comprehensive background in occult and theosophical lore. Based on these concepts, there are explanations of where devas (angels) and nature spirits originate, descriptions of their function in the scheme of evolution, and descriptions of their appearance. Finally, there are 25 color plates of water color paintings of these beings by an artist who worked closely with the author to assure faithfulness to his visions. The effect of these plates on myself was overwhelming. Each being is shown as a unique, human-like form surrounded and enveloped by swirls of colored energy showing the path of light through the aura. Some of the beings shown are mountain gods, landscape gods, and angel of music, a rose angel, the World Mother at the miracle of birth, and, most beautiful to me, three healing angels.

VIRGINIA BEACH VACATION: Last weekend, at the invitation of "Psychic Seekers" (a group to whom I lectured some months ago), I took my first vacation from ARF work—not counting the Florida visit last Christmas—which was a visit to parents combined with ARF work.

I did not really have to go to Virginia Beach for ARF, though there was a good talk by Charles Cayce there, and I saw the work of the Cayce Foundation and happened to contact a very important NASA scientist who may give us information on UFO s later. I also had the greatest pleasure of testing our Traveler's Directory by staying two nights at the home of a couple I had never met before (and having a great time there—just like old friends).

PHONE YOUR LEGISLATOR FREE: If you have a concern about any legislative matter, it is usually possible to speak to your senator, congressman, or to one of his staff free of charge. All you do is to call him at his office in Washington, person-to-person, collect. Generally, you will be told, "We cannot accept collect calls." You then reply that you know that, but that you also know that the legislator can have one of his staff call you back so that you can speak to him about your concern. They will then usually take your number and call you back within a day, especially if you say what you are calling about.

CANCER UNDERGROUND

Newsweek, September 17, 1973: The Cancer Underground:

To the recognized cancer expert, the tenth annual meeting of the International Association of Cancer Victims and Friends, Inc., last week offered about as much scientific enlightenment and much the same ambience—as a nineteenth century medicine show in the style of P.T. Barnum. But to the 1,200 mostly elderly men and women who crowded into the ballroom of the Los Angeles Marriott Hotel, the meeting was as spellbinding as a religious revival. Many of them cancer victims themselves, they paid seven dollars a head to hear about scores of bizarre remedies—ranging from apricot extracts to an electric enema—diligently touted by their producers as miracle treatments for this most dreaded of human diseases.

Founded by the Late Cecile Hoffman of San Diego, who had cancer, the IACVF has some 17,000 devoted members whose primary article of faith is that much of the surgery, radiation and drug therapy generally used in the treatment of malignant disease is harmful and "unnatural." Together with a tendency to adopt right-wing causes and religious fundamentalism, IACVF adherents also share the belief that "natural" and "nontoxic" cancer treatments are being kept from the public by a cabal of major drug companies, manufacturers of radiotherapy equipment and state and Federal health agencies. "It is all a matter of economics," declared IACVF board chairman Maurice Grant, a retired Phoenix lawyer. "A great percentage of surgery and radiation could be eliminated if nontoxic treatments were made available. But the medical profession doesn't want to give up the money."

But the drug that drew the most attention was Laetrile, an extract of apricot pits.

Introduced in the 1920's by the late Dr. Ernest T. Krebs of San Francisco, Laetrile purportedly destroys cancer by releasing cyanide once inside the malignant cell. After a series of investigations, both the U.S. Food and Drug Administration and California drug authorities found no merit to the claims made for Laetrile and banned its sale. But official action has done little do dim the enthusiasm of Laetrile's proponents, and supplies are smuggled into the U.S. regularly from Mexico and Canada. IACVF chairman Grant's wife testified at the meeting that Laetrile had cured her of breast cancer after surgery and radiation had failed. "All I know is I'm alive," she said.

But exotic drugs are not the only ways to fight cancer, according to IACVF speakers. Yoga was advised as a general health aid. And Paavo Airola, a 60-year-old nutritionist, argued that excessive animal proteins promote cancer and recommended that almonds, peanuts, buckweat and soybeans be substituted for meat in the diet.

The above excerpt from **Newsweek** attempts to ridicule Laetrile and other non-poisonous treatments for cancer. Insofar as I have been able to investigate this question, **Newsweek** is wrong and the "quacks" may be closer to the truth. I find that many people are healed of cancer whenever anything can be done which considerably improves the patient's general state of health. Therefore, it happens that so many people claim so many very different treatments. And they may be honest. It gradually becomes evident that cancer may not be a disease at all, but rather just a symptom of poor health.

Poor health cannot be cured by poisonous drugs, but it can be improved by all manner of good nutrition and good emotional attitudes. Laetrile (chemically, amygdalin) is actually a nutrient found in many foods which, however, have become rare in the western diet. It is found in millet (which was replaced by wheat that is more suitable to machine harvesting), sprouted mung beans, apple and other seeds and many fruit seed kernels. It does appear to kill cancerous cells in humans, and is non-toxic to healthy cells. But that is not sufficient to "cure cancer." Therefore it is called a control for cancer. To actually effect a cure, it seems necessary to overcome whatever else is causing the patient's poor health. Hence, so many "cures" that **do** work whenever they improve a person's health.

One of the serious questions about Latrile is its price: $3 per gram in Mexico, where large quantities are manufactured and labor is cheap. It is only about $2.35 a gram from Nutritional Biochemical in Cleveland, Ohio, which sells only small amounts for research purposes. Someone seems to be making a pile on Mexican Laetrile.

CABASIL, supplied as a food supplement by Dr. Stuart Kabnick, is another nontoxic food supplement which seems to be as effective in many "incurable" diseases as cancer, gangrene, etc., as Laetrile, or even more so, and we are investigating that. The daily requirement costs about 25 cents, as compared to $6 to $30 for Laetrile. To find out about Cabasil directly, contact Dr. Stuart Kabnick, Broad-Locust Building, Philadelphia, Pennsylvania 19102. (215) 735-8330.

PSYCHIC SURGERY is becoming better known, and more information on that field has been coming in since we last wrote about it. We have learned that whole planeloads of people are being transported to the Philippines from Europe, especially Switzerland, where a medical doctor who was suffering from serious heart disorders was cured by Tony Aqpaoa.

We have also learned that there is a psychic surgeon operating in the state of Georgia who does not want much publicity. He has a waiting list of several months or a year before people can see him. His methods are different from those of the Philippine surgeons in that he does not work on the physical body at all, but only on the spirit body in a manner very similar to medium George Chapman in England. People needing to contact him could get his address from me. They should probably read a book about Chapman's work first.

STANFORD RESEARCH INSTITUTE is one of the leading scientific establishments in America. I visited them at the request of Dr. William Kautz, who is interested in testing the work of Dr. Jonas in the field of astrological birth control. We were able to supply them with the original writings of Dr. Jonas and others in the field of scientific astrology, and we will be looking forward to hearing more from them.

HEALING THROUGH PRAYER: Can we learn it? In April, 1972, we reported on research in the field of prayer and its effects on the growth of plants, as measured by scientific plant measuring devices. Proof was offered that the rate of growth of grass was increased over 800% by the prayers of two well-known healers, Olga and Ambrose Worrall of Baltimore, Maryland. The plants were a thousand miles away when they prayed for their growth to increase.

We also told of research being conducted in New York to attempt to design an instrument that would measure a person's healing effect. We now have more information on that work.

To begin with, there is not sufficient evidence to indicate that the instrument in question actually measures a person's healing ability, and whatever it is that the instrument does measure may not even be related to healing ability. Nevertheless, the experiment is extremely simple, and perfectly safe so long as people do not use it for any kind of an ego trip.

The instrument in question actually measures the voltage which a person's body produces between the right and left hands, averaged over a ten second interval. It consists of nothing more than an aluminum can type electrolytic condenser of about 150 Mfd. (any voltage), a 0-100 (approximately) microampere meter, and a push-button to connect the meter across the condenser after ten seconds.

The claim is that a person holds the condenser in the right hand (the aluminum can is the negative lead of the condenser, and the hands are moistened with a salt solution), and the positive lead of the condenser is connected to a brass or copper electrode held in the left hand, then after a ten second interval a charge is found across the condenser, the condenser discharge causes the meter needle to rise up and return.

According to the information I have, different people produce a different amount of charge, or cause the needle of the meter to rise to a different point on the scale. There seems to be some correlation between the amount by which the meter rises and a person's mental or emotional state; that it may relate to a person's psychic healing ability, though, that has not

been proven. Considerably more research is needed in this matter.

I also understand that a person can learn to think in such a way as to cause an increase in the charge in the condenser, but I could not say whether to do so would be beneficial.

If there does turn out to be some correlation between healing ability and the potential difference between the hands, it would be very interesting to have a unit developed that could measure the variations in potential almost as fast as they occur, so that this could be more effectively used as a biofeedback device.

URI BY ANDRIJA PUHARICH

Each of us at ARF have now read the book by Dr. Puharich, M.D. We were so deeply impressed with the work and the experiences that Andrija has reported that we felt the need to telephone Andrija (whom we have known for a while) and ask to see him personally as soon as it is possible for him to talk with us.

In the book, Dr. Puharich sets forth the experiences in his own life which led him into full time research with Uri Geller. We learned in our call that, like ourselves, Andrija has ceased to work for money or to concern himself with financial security, and has given his full time to the work in trust that all else that is needful will be provided.

To summarize the book in a few words is not possible. Too many of the reported experiences are so uncanny when taken out of their full context that they cannot be believed, and if I had not myself spent hours carefully observing the things that Uri does (such as bending metals with non-physical means), I would hardly be in a position to accept the events Puharich reports. (I am still a skeptic at heart - Art.)

In **Uri**, Andrija reports many details of actual contacts with beings that give evidence of extra-terrestrial origin; he tells how he was forced to accept this evidence by the physical things that happened in his own and other people's presence, also including the uncanny telepathic and mind over matter feats which Uri continues to perform, and which have been investigated at the Stanford Research Institute (I have also seen the SRI film and news release, both of which are reported in the book).

In a final effort to summarize the book (at least in part) so that you can better decide if you want to read it, Bruce Dixon reports:

"The book includes a scattering of facts (?!) (they are so uncanny, that I understand Bruce's question mark well) about the particular extra-terrestrial brothers. They call themselves the Hoova race, some or all of whom live on a gigantic space craft the size of a small planet called Spectra. They claim to have first contacted the human race 20,000 years ago (is that figure correct?) at the 'Oak of Mamre' in Hebron, where Abraham met them. The biblical reference is to a 'ladder of the gods.' They give advice to man . . . and were an active force in nurturing civilizations in Egypt under Imhotep, again in Aztec times under Tehouti, and in Alaska, China, and India. They make no claims, however, to be the sole creators of any culture.

"Also, they affirm the existence of a soul and the experience, at least for some, of reincarnation into human bodies. They themselves apparently have astral/etheric, non-physical bodies which endure for a million earth years. It seems that in ways not yet revealed, the Hoova are bound to us. They say that they wish to land their space craft near Earth in order to enter Earth's fields and 'recharge.' This event may, or may not be visible to humans, depending on our ability to accept it naturally and calmly."

In his discussion of the book **Uri**, Bruce goes further and compares it with a little work called **Links With Space**, put out by the Findhorn Community. I have not read this yet but Bruce writes:

"**Links With Space** is a collection of three transmissions from 'angelic' sources received at Findhorn around 1970, and contains an excellent introduction by David Spangler (whom we have met). Some of the information here is that 'Space Brothers' were here during the 60's, partly for the purpose of evacuating earth people who might be builders of a new world in case there were an atomic war, which would have destroyed our present civilization. Fortunately, they believe, that danger is now past.

"Since the late '60's, many sensitive people have begun to feel an increase in stability and creativity on the planet, and instead of waiting apprehensively for the fulfillment of prophecies of doom, they have become active in seeking a new plan and a new vision for mankind. This action, taken by small but influential groups such as Findhorn, has caused 'enough light to be anchored' so that fear of destruction is no longer necessary.

"The introduction gives crystal clarity to the confusion of predictions of destruction and simultaneous predictions of a New Age. It must be read in its entirety to be understood.

"Because of these changes for the better, space contacts on the physical plane have declined, but extra-terrestrials have not ceased contact. They have simply shifted to anotner form of contact. It seems that in this form, there is actually more possibility of contact with us. . . . It's claimed that free commerce and communications with other planets in cosmic brotherhood is natural and inevitable for man when he becomes equal to the challenge."

Uri is published by Doubleday's Anchor Press and sells for $7.95. Perhaps it will be out in paper soon.

P.S. There are one or two passages in **Uri** which seems to me to confirm my ideas of the future of humanity, and the main purpose of our visit with Dr. Puharich could be to see to what extent he feels those ideas to be in line with his experiences and information from the sources mentioned. Art R.

EXECUTIVE ESP by Dean, Mihalaski, Ostrander & Schroeder. (Prentice-Hall: $7.95)

This is a book we have been waiting for. It is a well written account of modern psychic research in which two pioneer scientists from Newark College of Engineering have teamed up with the authors of **Psychic Discoveries Behind the Iron Curtain**, and produced a most readable, yet scientific report of American research in the field of ESP. Actually, I've only read the first 83 pages so far, but I can see that it is far more factual than other works of Ostrander and Schroeder, due mainly to the cooperation of good writers and good researchers working together.

Executive ESP is a book that I will have to work through and write up more fully in the future. It will not sweep the nation as quickly as **Psychic Discoveries** because it is a more ponderous, scientific work, but it is bound to affect us all profoundly in the next few years. Read it. (Note: The book is 200 pages in length, but added to that are about 90 pages of appendices (many), index and bibliography.)

One of the most mind-blowing chapters of **Executive ESP** that I came to only tonight is entitled: SEXUAL DOMINANCE: HOW IT AFFECTS YOUR ESP.

In their research, Dean and Mihalaski discovered that a person's precognitive ability could be measured by allowing the person to guess a series of 100 numbers that a randomly programmed computer would make up in the future. By chance alone, one would be expected to guess 10 of these numbers correctly. When a person consistently scored above 10, this was an indication of precognitive ability.

It was discovered that a person's precognitive ability (and therefore, presumably their general intuitive ability) declined sharply when they were in a situation which placed them in a minority situation in any important aspect of their lives. For example, women functioned poorly in a male dominated situation, and men, likewise functioned less effectively in a situation in which females were dominant. This effect was most pronounced in people who would be classed as the "dynamic" personality.

It seems evident then, that in order for groups such as our own little community to function efficiently it is important that there be as little dominance by one sex or type of person as possible, and that it is more necessary to have one or two additional women in the group than feelings alone would have indicated.

NEW AGE FOCUS: SYNERGY ACCESS

Synergy Access is an 8-page bimonthly paper put out by Wes Thomas & others in East Northport, New York. Issue #5 is directed towards New Age thinking and has an update on the work of Dr. Puharich & Uri (following publication of their book) as well as many other New Age items.

S/A specializes in giving a short indication of many different sources of information; always giving names and addresses and the concern of the authors is very similar to that of ARF. The paper is especially important for people who work with the media in any form.

A quote carried in #5 is very important and expresses very much the concern of ARF; it is taken from the words of a recently deceased leader of the Auroville community in India who is known as "The Mother." She said:

"The future of the earth depends on a change of consciousness and the change is bound to come. But it is left to men to decide if they will collaborate for this change or if it will have to be enforced upon them by the power of crashing circumstances."

Evidently, the Mother saw the vision that we also see. If enough of us work hard at helping the New Age to come, it can arrive quickly and peacefully, the old order descending as in a parachute. Otherwise, it will come eventually, but only after suffering and violence such as pictured in Revelations the last book of the New Testament.

To really learn what **S/A** is about, I suggest that you send $.50 to **Synergy Access**, 606 5th Ave., E. Northport, New York, 11731 (phone: 516-368-2609 anytime). Ask for a sample copy of issue #5. Subscription is $5 per year.

INST. FOR LOCAL SELF-RELIANCE This is a group of people in Washington, D.C. (1717 18th St., N.W., 20009) who are working to provide ways in which the bigness of our society can be overcome, and even city dwellers can develop some of their own food production, services, & even electric power supplies.

Most of their staff seem to be professionals on the scientific levels, Ph.D.s or better, or former government or law specialists.

ASTRONAUT MITCHELL SLOWED BY FINANCIAL DIFFICULTIES The Institute of Noetic Sciences, directed by former astronaut Edgar Mitchell, appears to be having difficulties in financing the programs originally intended. They write:

". . . therefore we have had to reduce staff at the Institute, with a consequent reduction in programs as well. We are trying to continue with our mission, but inevitably we will have to give deep reconsideration to our goals. If we cannot reasonably carry on activities in some area, we will simply have to withdraw from it.

"This is a time of testing for those who are committed to helping bring forth a better society and a New Age. We are considering a reorganization of the Institute, and will keep you informed as major decisions are made. In the meantime, your continued support is — as always — gratefully received".

The June newsletter of the Institute then tells of a 700 page book by Edgar Mitchell & Associates about to be published: PSYCHIC EXPLORATIONS: Price - $17.50 ($15 to Institute members.) Tapes of a lecture series are also offered at $6 each. Address, Institute of Noetic Sciences, 575 Middlefield Road, Palo Alto, California 94301.

Just as many New Age groups and communities are moving into new and deeper visions, Noetic reports decline. When Mr. Mitchell was in Philadelphia, honored at an expensive banquet, I was told that there was no chance to speak with him, for his schedule was arranged and filled weeks in advance. I would have liked to ask him what he felt the New Age would be like, and whether it would mean highly paid employees or brothers and sisters working together in love. It seems to us that these times have a message for us that suggests that to work for the New Age it is important that those who do the work are living together in the spirit of the age that they seek. We hope that the re-organization which Noetics forsees will lead them into the close harmony and dedication of sharing brotherhood so that they can go forward with money or without it.

HYPNOSIS COURSES Beatrice Clegg, hypnotherapist, consultant and lecturer, is helping many people overcome their limitations through hypnosis. In a published lecture entitled, "Intercommunication for the Dying and Bereaved," Mrs. Clegg lists many successful cases using hypnosis.

At a group meeting in a Philadelphia hospital, we demonstrated muscle responses of a hypnotized subject to suggestion. Our "silent witness" was a research model electromyograph (EMG). Readings were taken of the left and right shoulder areas. Measuremenets displayed on the oscilloscope of the EMG indicated both shoulders read the same.

"*I hypnotized our clergyman subject with suggestions to reproduce a discomfort in the left shoulder area similar to that which he experiences from time to time (but was not present at the beginning of this demonstration). The EMG showed a dramatic increase in muscle irritability in the left shoulder only. As relief was suggested, the EMG oscilloscope reading followed suit until the left and right measurements were again the same.* Fascinating, yes?

"In 1961, I worked as a hypnogenic consultant in a rehabilitation facility with people who had injury in the brain or spinal cord areas. Hypnosis there was useful in a myriad of ways and with varying degrees of success. J., a severely brain injured 17-year-old girl, victim of a tragic automobile accident, developed an external body fungus. There was doubt as to her comprehension. She was withdrawn and uncooperative. *The Medical Director asked me to explore her hypnotizability with the idea of curbing her perspiration through the sympathetic nervous system. Medication would be continued.*

"*J. was hypnotized at our first session and rapidly became more cooperative in the various therapies. She began to take an interest in looking at fashion magazine pictures with me. It was not long before the Medical Director announced at staff meeting that the fungus was drying up much faster than any he had seen with the use of medication alone.*

"*Was J.'s perspiration slowed? Or was healing stimulated by suggestion with no effect upon her sympathetic nervous system? Or were both factors involved? I suspect the latter.*

"*There are reports that for 18 months, a highly regarded radiation therapist has documented many cases of cancer patients whom he has taught to use meditation combined with treatment. He states his patients do much better with meditation than without it. His techniques, as described to me, are similar to what for years I have been teaching in self hypnosis pertaining to various illnesses and when the patient's physician deems it advisable. These cases, too, are well documented. The semantics here are not important–but the results are.*"

LOW COST NATURAL HEALTH CLINIC: In the last week, I have had the opportunity to visit a unique low cost natural health clinic in Quakertown, Pennsylvania. The clinic is very unlike the city places where people are treated as numbers. Instead, it is a small group of buildings in a very rural setting three miles outside of Quakertown, and 40 miles from Philadelphia.

The doctor in charge is G.E. Poesnecker, who was formally trained as a chiropracter, but whose knowledge ranges far and wide into many healing fields, so that many types of affliction are effectively dealt with, mostly by natural methods. A health food store and eating place are part of the complex. Dr. Poesnecker seems to be always busy and active, yet very cheerful and relaxed.

Fees at the clinic are also low; a visit and consultation with the doctor is $7. One dollar extra is charged for treatment of 20 minutes by a $5000 radiation device (similar to a dia-pulse). Medicines, mostly of the homeopathic variety cost about $2, usually, and are supplied at the clinic. Other doctors, also of the more accepted AMA type, cooperate with the work at least on a part-time basis.

I was impressed by Dr. Poesnecker's openness to new methods; within a week after I had presented him with the information on cancer which is presented here, he had spoken with Dr. Gold and applied for proper legal permission to use the drug in an experimental capacity for one of his patients.

NINE SHEPHERD PUPPIES were born to Lady while I was in California, which was a week before they were expected. I had planned to get home before the pups came, but their premature arrival made it possible for me to make the trip more leisurely. It was also a strange experience for me, as I woke up in the morning having dreamed about Lady having puppies. I called home that she had had them the night before. I had not previously dreamed about her at any time.

VISIT WITH PETER CADDY AT LINDISFARNE Lindisfarne looks like a group of 19 summer cottages with a main building and dining hall. In reality, it is a New Age learning community headed by William Erwin Thompson. In May, Lindisfarne was host to Peter Caddy of Findhorn on the first stop in his initial American tour. We made arrangements, to drive up and meet Peter with Elizabeth Faust.

Lindisfarne is located on Fishcove Road in Southampton, New York 11968, and their phone number is (516) 283-8210.

Interviewing Peter was mainly meeting a man with a lot of peace to share. He told us that two persons claiming the ability to see nature spirits once reported seeing the same thing upon separate observations, but a serious scientific study intended to verify nature spirits has never been considered.

Peter gave his opinion that we need no longer resist or fight the establishment. Energies are being withdrawn from this cilivilzation; money is being given to alternative projects, people are investing in rural land, and persons on the top and bottom of society are peaking out into more fulfilling activities. The best we can do now, he says, is to first find a sense of peace within ourselves, then build peace within a community, and then peace between like communities. Peter adds that we are all involved in giving birth to a better world. Sure, there will be birth pains, but we must not concentrate on the pain, but on the beauty of the child who is coming.

Findhorn's nearest neighbor is an R.A.F. base. Peter was in the R.A.F. during WW II, and there is a fruitful exchange between them. In fact, an officer after touring Findhorn and sitting in the sanctuary has decided to work for them and will be employed as an administrator.

We saw an impeccably produced slide and tape show that revealed Findhorn, not as the large commune we had envisioned, but as a physically small operation, perhaps 3 acres. Their real strength is not in size, but as a potent center of demonstration to the world of fulfillment in co-operation, creativity, and spirituality.
Creative Education." The University intends to be a training ground for persons to become the leaders in the New Age.

Peter hopes to make more tours and we have invited him to Philadelphia, but he made no promises.

PREGNANCY A PROBLEM?
The Farm: We recently reported on a visit to the communal farm of Stephen Gaskin in Tennessee.

While there, I was told that the group's outlook was strongly opposed to abortion and that I could publish the fact that any woman who is pregnant with a child that she is not able to care for could feel welcome to come to the farm. She could have the child and leave it there, and it would be cared for until such time as she might want to have it. No doubt other spiritual communities would agree to do that also, but in this instance, I was told that I could freely publish the information.

Contact The Farm, Rt #1, Box 197-A, Summertown, Tennessee 38483 (615) 379-7246

THE NEW HUMANITY As the final word in the book, I would like to share with you a series of experiences which, I feel, have led me to a deeper understanding of the message of Jesus of Nazareth, and the prophets who went before him. After living in religious communities for sixteen years (such as the Society of Brothers at Rifton, New York, and The Reba Place Fellowship in Evanston, Illinois), I began to work with people who were putting their full time into stopping the war in Vietnam, and this led me to a National Council meeting of SDS in Berkeley, California in 1966.

One evening I found myself together with an intimate group of 25 or 30 people of all ages and sexes when it was suggested that each one should speak for about two minutes to tell how he would like the world to be in the future. At the time, we were sitting in a circle on the floor and the only light came from candles held by each one of us.

Frankly, I was amazed at what came out; not by what any of the people said, but by the fact that all of us seemed to be in total agreement with every single desire that was expressed in that circle. In all the years I had lived in communal societies, I had never seen such complete unity on so many different points so readily arrived at, and unity had always been the community's primary concern. After wondering about this for some time, I theorized that the reason for such unity that evening was not the nature or condition of the group, but more likely the subject of our concern. In telling how we wanted the world to be in the future, we were actually describing the deepest longings of all humanity and our complete unity could only imply that the basic longings of all human beings might be exactly the same.

I thought about that for some time, and then tried to design an experiment which could indicate if that were so.

I was working at a place where people would come to visit from all over the United States and from other countries, so when I had time I would tell the visitor that I was not able to tell his fortune, but that I could tell him what his deepest longings were and that I was usually about 95% correct. There was no problem finding candidates; everyone wanted to have the 'reading." I would always warn the subject that some of what came out might be a little embarrassing, for I would probably detect some things that he had never before confided to any other person. People generally asked to have the reading done in private.

I always started by examining both sides of a person's hands, and then I explained that I also had to look behind their ears, and I examined that area quite carefully.

The subject generally supposed that I was, in some mysterious way, able to detect his own deepest longings, and that he was receiving an individual, personal reading. A typical reading might have proceeded as follows:

"Oh, wow! I see that you have a longing for tremendous freedom." (Pause; subject agrees.)

"You want to be so free that you could just flap your arms and fly like a bird! I see more: you even want to be able to fly right through space to distant planets and explore and meet people there. Fantastic!"

"It seems that you also have a great longing for all kinds of adventures, but you want to have adventures for a purpose; not just adventure for its own sake. At the same time, though, you also seem to have a longing for security that you will be able to live forever and never have to get old or senile. You want to be completely free from all illness, and you also want to be able to cure other people quickly and easily."

"You also want to live in a world that is full of peace and harmony and love and joy, and you would like to be free to have children without having to worry about money or economics. You would like everyone to share with everyone else, so that if you wished to travel anywhere you could be sure that the children would be cared for just as if you were there. Everyone would love all the children, too."

(At this point I might study the ears or hands a bit more, noting some peculiarity.)

"I see that you also long for the freedom to be completely open and honest with other people; you want to be fully understood by them, and want to understand them fully also. Mmm, look here, it goes even further; you want to be able to talk with animals, and even with insects and trees and plants and have similar rapport with them!"

"Oh, now I have to meditate for a moment. I see some things here which may be very embarrassing, and I'm not sure if I should speak of them." (Subject promises not to get upset and begs me to tell all. I agree to comply with their wishes.)

"You have a longing for tremendous freedom, also with respect to your own body. You would like to be so free that you could be naked together with other people of both sexes and not be ashamed at all." (Actually, that was the state we were in at the meeting in Berkeley, and therefore it was possible for people to speak more freely of this than they are usually prepared to do.)

"You even want to be free to have sexual relations with others, and not be restricted to a single partner; you would like to be able to be completely open and honest about such relationships, and to be completely free of jealousy or envy. In fact, it seems that you would like to feel free to be physi-

cally affectionate with people of your own sex, also, and to feel no shame about that." (Subject smiles, but agrees; insists that he never breathed a word of this to anyone and is not homosexual, but would like freedom to embrace a fellow when he feels like it; how did I find it out? I promise to tell him that later.

There are probably a number of other longings I could tell about. With a bit of thought you could guess what they might be. Yet it almost seems weird to tell of these so quickly and easily. How is it possible?

Apparently, each person's longings are almost identical with those of other people, but the culture in which we live has discouraged us from being sufficiently free with ourselves and each other to admit this reality. I only have to know what my own longings are and I am able to tell you what yours are; only a few slight differences between the sexes, such as a woman longs to have a baby suckle at her breast and things like that. All this looking at your hands and ears was only a trick to make you feel that I was giving you a really personal psychic reading so that you would more readily admit some of the things that we are usually too shy to speak of. We live in a strange society when people don't dare to admit things which are really the same for all of us. I hardly know you, and that I can tell you all this is proof that it is the same for all of us."

After confirming this many times, even with people from India and Japan, I became convinced that these longings were most likely universal, and I began to wonder why this was so. I thought again of Jesus and the prophets who spoke so often of a resurrection; a time to come when all who have ever lived would live once more with new and more perfect physical bodies. Then I recalled some of what Jesus had said and done. I remembered things like: "Be ye perfect," or "Greater things than these shall ye also do!" (which He said while referring to the miracles). I remembered how in the end he was seen to fly up into the clouds as we long to be able to do, and of course all his examples of healing and love. "Love one another as I have loved you!" I also recalled the evidence that his love was physical as well as spiritual, for we have the description of the Lord's Supper where John was laying with his head on Jesus' breast. There were so many other examples that there is not room to mention them. He always said "Repent for the kingdom of God is at hand." (Turn around for the rulership of love is coming.)

But there was one thing that troubled me a lot. At one point it is told, the Sadducees came to Jesus and asked him what would happen if a woman had been married to seven husbands, all of whom had died; when the resurrection comes, which one would be her husband? (The Sadducees were a sect that did not believe in a resurrection.) Jesus had replied:

"When the resurrection comes there will be neither marriage, nor giving in marriage, but ye will all be as angels."

This, to me, was painful. I knew that I should look forward to the coming of God's kingdom, but I felt sad to think that I might never experience the joy of physical union which, in our community, could only take place in marriage. I wondered, also, at the redundant wording. If there would be no marriage, how could there be "giving in marriage"?

That was how I had seen these things when I first learned about them in a Society of Brothers community in Paraguay. Now it was a few years later, and I began to wonder how it was that all of our longings seemed to fit what Jesus had said and done except in the area of marital love. I didn't know about angels, but I wanted to love and be loved. Why did Jesus exclude that?

I thought first about "giving in marriage," and was aware that in his day, women were given by their fathers to whomever the father chose. They had no choice; they were as slaves. If they 'committed adultery," they were to be stoned, but there was never any mention of a man being punished for adultery. Perhaps when Jesus said, there would no longer be giving in marriage, he was meaning that the slavery of women would be ended.

And if that were so, what about the "bond" of marriage? Envision a time when everyone would be open and honest with one another, and there would be no jealousy or selfishness; no money or private property, no death, disease, or suffering. Everyone would love one another and certainly all the children. In such a time, what need would there be for marriage?

So now it appeared that the deepest longings of mankind could be the same as the greatest teaching of the masters, and that we have these longings because they are a picture of the future; a clear vision of the new humanity that is to come for all of us. I came to feel that in our present bodies we are like babies in a crib; a soul in a body.

Afterwards, we may pass on to a higher school of spiritual learning, and finally, when the time is ripe, we are to come together in perfection with the new bodies the Bible referred to, and which Jesus was said to have had after His resurrection.

If this were reality, then the longings which we seem to have in common make perfect sense, and that might be why so many people are beginning to feel that man's consciousness is rising. People seem to be less concerned with "roles" and more with friendship and love, and that we are indeed approaching a turning point in human evolution. Perhaps many of you can try the experiment which has been described and see if it also works for you and let us know.

If it is true that the life of each individual is a permanent reality, then we may have chanced upon the main cause of all the cruelty and suffering presently afflicting mankind. Most people are living in the illusion that their human life is only temporary and so they feel the need to try to fulfill as many of their natural longings as possible and they try to do that quickly, before they grow old and "die." As a result, we have all manner of selfishness leading to capitalism, communism and war. When we become certain that our lives are really permanent, and that all of our longings will ultimately be fulfilled even if we must first pass through a change of state called "death," then we will no longer want to trample over our fellow humans and destroy our environment to achieve our own selfish ends.

Instead, we will realize that the true humanity must come about for all of us together, and to achieve it we must work for the whole of mankind rather than for ourselves or our group, alone.

For the New Age to come about gently and speedily, many miracles will have to occur, and we don't understand how they can come about. Yet, we have already experienced so many surprises of a miraculous nature (as the reader of UNPOPULAR SCIENCE is aware) that we should soon be able to trust that if we do whatever we can, other persons and powers will help with the rest.

These newsletters, and the whole ARF project is our attempt to be open to this and we are eagerly looking forward to a great deal of additional help from many of you. May the New Age of love and joy come a running!

Love & peace,
Art Rosenblum